IMAGES
of America

EASTPORT

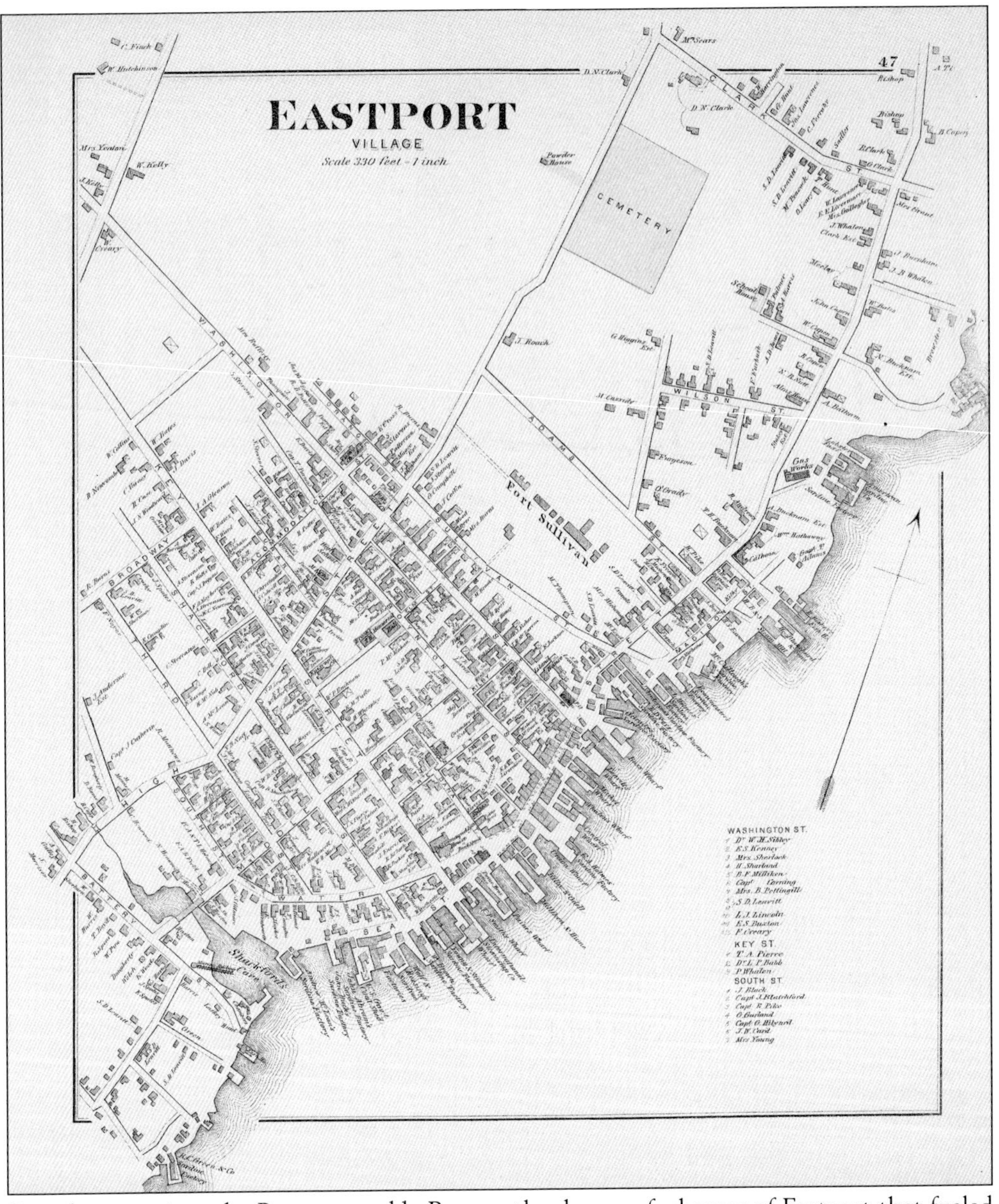

Stretching out into the Passamaquoddy Bay are the dozens of wharves of Eastport that fueled the packing and shipping of more than 13,000 cases of sardines every week; this map is from the 1881 Washington County Atlas. The small island city gave birth to the North American sardine canning industry. Atop the hill just above the center of the map sits Fort Sullivan, a military complex once seized by the British during the island's occupation from 1814 to 1818. (Courtesy of the Tides Institute & Museum of Art.)

On the Cover: Artists and island life have long gone hand in hand in Eastport. In the 1920s, this group of plein air painters are in the midst of capturing the visage of a maiden clad in white with Bungee Wentworth's maritime general store behind her. The store was located by the area at the head of Shackford Cove, colloquially known as "Sodom" for the challenging and dangerous living conditions its residents lived in. (Courtesy of the Tides Institute & Museum of Art.)

IMAGES
of America

EASTPORT

Lura Jackson with the
Tides Institute & Museum of Art
Foreword by Hugh French

ISBN 978-1-4671-0745-7

Published by Arcadia Publishing
Charleston, South Carolina

Printed in the United States of America

Library of Congress Control Number: 2021946409

For all general information, please contact Arcadia Publishing:
Telephone 843-853-2070
Fax 843-853-0044
E-mail sales@arcadiapublishing.com
For customer service and orders:
Toll-Free 1-888-313-2665

Visit us on the Internet at www.arcadiapublishing.com

For the people of Eastport—past, present, and future; for my family; and for those who would know the island at the edge of the world. Together, we will keep the light of learning lit.

Contents

Foreword

This book adds another important layer to the understanding of the life and history of what is now the very small city of Eastport, Maine. Located on the northeast corner of what is now the United States bordering what is now Canada, this place lies deep within traditional Passamaquoddy territory that encompasses both sides of the current US/Canada political boundary.

Much of what is known of the history of this place is the culmination of many efforts over time. Lorenzo Sabine, an early-19th-century historian, collected large amounts of original materials and wrote from these. Jonathan Weston, a lawyer, presented the first history of Eastport in 1834. William Henry Kilby compiled a series of chapters in book form published in 1888. Historical articles in newspapers largely took over from here, first within the *Eastport Sentinel* until it ceased in 1953 and then the *Quoddy Tides* from 1968 through today. The Peavey Memorial Library long served as a repository for local historical items within its building, constructed in 1893.

During the 20th century, when Eastport experienced significant population decline, antique dealers set up shop in downtown and took truckload after truckload of antiques from this place. As a result, much of what one might have expected to find here was no longer here. To counter this loss, the Border Historical Society began in 1963. Local historian John Holt wrote *The Island City* for Eastport's 1998 bicentennial, and another historian, Wayne Wilcox, has researched the area for years, with much of his findings published in the *Quoddy Tides*. Professional historians have also written recently about Eastport history. The Facebook group Old Pictures of Eastport, Maine, provides an invaluable service to virtually connect this community to its history. Donald Soctomah, historic preservation officer for the Passamaquoddy tribe, has spent decades painstakingly researching and sharing its history. The Tides Institute & Museum of Art, founded in 2002, has amassed the most significant collections relating to the history of Eastport and the surrounding area, securing artifacts long missing from the area as well as those still here, including much of the collections transferred from the historical society and items from the library as well.

One hopes that more of Eastport's history will be discovered in the future and that a fuller picture of its history can be formed.

—Hugh French

Acknowledgments

First and foremost, thank you to my parents, Steve and Judy Crawford, for helping me feel comfortable with exploration—and for the spins in Old Sow. Thank you to my sister Tobey for being among my first companions.

To my partner and muse, John: Thank you for all the threads you weave and see, spinning light and voice into embodied humanity.

This book involved the efforts of several people. Hugh French was highly instrumental in scouring the Tides Institute's archives and providing appropriate photographs and in reviewing the final draft. Wayne Wilcox and Susan Esposito provided invaluable help in proofreading and spotting errors in the early drafts. Al Churchill of the St. Croix Historical Society provided insights and materials. Each of you has my eternal gratitude for your work in preserving and promoting the area's history.

The Eastport community is graced with many talented photographers, and some of their work appears in this book. Thanks to Don Dunbar, Edward French, Robin Hadlock Seeley, and Tom McLaughlin for providing photographs from their personal collection.

The entrepreneurial families of Eastport are second to none in their support for community efforts. Thank you to Karen Raye for photographs of Raye's Mustard Mill and Paula Bouchard for the information on the history of Rosie's Hot Dogs.

Lastly, thank you, dear reader, for joining me on this pictorial tour as we continue to turn each page together.

Unless otherwise noted, all photographs in this book are from the collections of the Tides Institute & Museum of Art.

INTRODUCTION

If you are traveling both north and east in America, and you do not stop, one of the places you will likely end up is Eastport. Located on an island joined to the mainland by a causeway, this "city" is the smallest municipality by area in Maine. Compensating for its lack of size, however, is a wealth of character—people who have been to Eastport are not soon to forget it.

It is hard to put a finger on what exactly makes Eastport so unique. The Northeast is filled with harbor towns, each hawking its own seaside wares and ocean fare. But it is Eastport alone that offers the deepest port on the East Coast, with an accompanying topographical layout that produces truly dramatic ocean effects.

Not far offshore in the Bay of Fundy, the ocean depths plunge to hundreds of feet, while nearby channels reach just over a few dozen meters. The result is the perpetual cadence of tides coming in followed by tides going out—shifting as much as 26 feet in six hours' time.

With so much deep water surrounding Eastport, and the water always in flux between high tide and low tide, the oceanic energy around the port city is significant. With that in mind, it comes as little surprise to find Old Sow, the largest whirlpool in the western hemisphere, swirling endlessly near the shore.

While the ocean is a defining part of Eastport's story, it mainly serves to provide a backdrop to the human drama that has unfolded—and continues to unfold—within and around its boundaries.

The island now known as Eastport has gone through many iterations in its time. Initially carved by the weight of the Laurentide ice sheet as it receded over the course of 10,000–15,000 years ago, the mostly granite outcropping found new life with the subsequent emergence of evergreen forests, marshes, and wildlife.

Soon after the ice departed, humans arrived. Crafting lightweight canoes from birchbark, durable tools from flint, and formidable weapons to take down large game on land and at sea, the Passamaquoddy people became a part of Eastport and its surrounding lands and waterways.

Named for their technique of hunting pollock with a spear (*Peskotomuhkati* in their native language), the Passamaquoddy flourished across the area that would become eastern Maine and western New Brunswick, Canada, for the next several thousand years. Consisting of 20,000 people prior to European contact, the tribe utilized well-established camping sites along the coast and inland depending on the seasons.

Directly adjacent to Eastport is Sipayik, or Pleasant Point, a former peninsula that once served as the ancestral Passamaquoddy home during the warmer months. With direct access to the ocean, the Skutik (St. Croix) River, and abundant mudflats, the location is ideal for harvesting fish, shellfish, eggs, and marine mammals.

As winter approached, the tribe would pack up its village and travel inland along the Skutik River until it reached Motahkomikuk (Indian Township), a lakeside area that was perfectly suited for weathering out the harsh cold of the northern wilds.

Following the arrival of European explorers in the area in the early 1600s, Eastport, Sipayik, and the surrounding environs became a place of trade between the French and the Passamaquoddy, with the British sometimes joining in (though the British were markedly more hostile to both parties). Early exchanges primarily involved the luxurious coats of the otter, mink, beaver, and their cousins for metal goods (such as tools and guns), though the Passamaquoddy were prolific traders who would have had occasional exotic wares from farther south along the vast indigenous trading network.

Sadly, physical goods were not the only things exchanged, and a proliferation of disease and violence killed more than three quarters of the Passamaquoddy people within a few short years.

Long accustomed to the drumming, chanting, and dancing of the Passamaquoddy, the waters and lands of Eastport were soon home to the sounds and sensations of Western civilization. Permanent frame houses cropped up across the island, followed by docks, smokehouses, and an ensuing flow of sailing ships.

While settled by a few dozen patriotic American families in the late 1700s, Eastport's proximity to English territory (with the soon-to-be-Canada directly visible across the bay) made it uncomfortably contested for the first several decades of its incorporation. The British felt strongly enough about the situation, in fact, to bring a fleet in to capture the island during the War of 1812. The four years they remained in Eastport make it the most recent area to have been occupied by a foreign power in the contiguous United States.

After gaining its freedom and Maine soon gaining statehood, Eastport began a slow ascent into becoming a booming coastal port. In 1876, the town was propelled to new heights when it saw the opening of a sardine factory that heralded the very beginning of the massive North American fish canning industry.

Eastport, as a land and as a city, took on a new identity. On one side of it were the wealthy merchants, captains, and retired military families, many of whom enjoyed lives enriched by the heights of Victorian culture. On the other side were the dock workers, fish cutters, deckhands, and farmers who were keeping the city's industry going through long days and nights spent in challenging conditions.

Despite the dichotomy of the Victorian era, Eastport retained a sense of distinct cohesion. No matter if you were traveling to Boston on a steamship in first class or in the boiler room, if you called Eastport home, you were part of an extended island family. The city's small size (measuring four and a half miles long and not wider than a mile and a quarter at any point, according to William Henry Kilby's *Eastport and Passamaquoddy*) and limited street layout—and the resulting compression of the community—is one reason its residents tend to be closer than those of other municipalities.

For the Victorians, another factor lending to the sense of solidarity between Eastport residents was the continual procession of major disruptions (with the British invasion in 1814, the Saxby Gale striking in 1869, and a fire devastating the downtown in 1886, to name a few). Time and time again, the community was tasked with banding together its resources to ensure its shared survival, and the overall effect was the creation of a layered, lasting bond.

As the sardine era waxed, sprawling factories took root on the shore and stretched out into the ocean over the steadily changing tides. Millions upon millions of cans of sardines were shipped out of Eastport as it became one of the busiest ports on the entire East Coast.

Like the fur and whale trades before it, however, the sardine industry could not last. The speed and scale of the depletion of the natural stocks was far too fast for the fishery to maintain itself, and the collapse took a portion of Eastport's prosperity with it.

The loss of the sardine industry was not the end of Eastport's story—far from it, in fact. The island's unique location, deepwater port, and access to natural resources have granted it a lasting fortitude as a place of interest for industries and artists alike.

In the 1920s, for instance, Eastport's massive tides and the energetic potential of harnessing the flow of the surrounding 110-mile Passamaquoddy Bay made it a prime spot to build the world's first tidal-powered hydroelectric dam. The project became a darling of Pres. Franklin D. Roosevelt, who himself had spent his youth sailing around the area's pristine waters.

While the Passamaquoddy Tidal Power Project was never fully constructed due to a funding shortfall stemming from Congress, the 25-year effort to build it (and the 5,000 workers it called for) left Eastport with a whole new residential area. Quoddy Village endures today as a remnant of the potency of the dreams the Eastport area inspires.

And though the Quoddy Dam project was abandoned, efforts still continue to develop a viable tidal power project that could feasibly power a significant portion of the state from Eastport's shores.

The fisheries and aquaculture industries have long been the lifeblood of the Eastport area, and while the species being harvested and farmed have changed, the island's strong connection to its fishing roots have not. Lobster, scallops, clams, salmon—and even seaweed and bloodworms—have all been nurtured and collected there, producing a proud legacy of fisherfolk who have called the area home for generations.

Alongside the rich fishing tradition of Eastport stands an equally tall pillar of artistic productivity. For well over a century, the island has attracted artists of all persuasions, drawn to the uncommon beauty juxtaposed with coastal industry. Fostered in part by the Tides Institute & Museum of Art, the Commons, the Eastport Arts Center, Stage East, and the Eastport Gallery, the artistic culture of Eastport is a thriving homage to the creative passions the island creates.

Music aficionados who are unfamiliar with the Passamaquoddy Bay Symphony Orchestra are continually awed by the talent encompassed in its performances, which are guided by Norwegian-born director Trond Saeverud and accomplished pianist Gregory Biss. Eastport Strings, meanwhile, offers a haven for string enthusiasts under the tutelage of Alice St. Clair.

Amidst its natural beauty and artistic wealth, however, the people of Eastport continue to face economic challenges. Part of the sacrifice of living in a rural refuge like Down East Maine is not having access to well-paying, white-collar jobs, and Eastport has long felt that burden. The economic disparity and distress in the region have driven some residents away, while others, unable to leave, have struggled with prescription opiates and narcotic addiction.

Like much of rural America, the lack of high-quality jobs has caused the youth to take flight to more prosperous regions; some of them later return to raise their families next to familiar waters. Though steadily dwindling in population, the local high school offers occasionally robust science, artistic, and athletic programs that gain state acclaim.

The Passamaquoddy tribe remains a powerful voice in the region, with Sipayik and Motahkomikuk (Pleasant Point and Indian Township, respectively) established as reservations that continue to honor their proud indigenous heritage with abundant youth and community programs. Every year at Sipayik, the Indian Day celebration in early August includes ceremonial displays and traditional dancing. Road access to Eastport is via a causeway running through Sipayik, along with drinking water access via the Passamaquoddy Water District infrastructure.

In its newest iteration, Eastport has opened its arms to offer itself as a wholly unique place to visit. The city hosts the Salmon and Seafood Festival foodie event every late summer, followed by the largest pirate festival in New England the following week. The airport is expanding, with day flights to Boston anticipated in the near future. Celebrities, politicians, and film crews have been attracted by the island city's efforts to perpetually renew itself.

The story of Eastport is as winding as its shores, but the constancy of its existence as a unique island community is as solid as the granite it was constructed upon.

One

Eastport in the Indigenous Era

The first humans to come to know Eastport did so in the wake of the massive Laurentide ice sheet, a two-mile-deep glacial body that began to retreat to the North Pole 20,000 years ago. As it receded, a landscape of granite, limestone, and slate was revealed in the form of islands, cliffs, and mountains.

Over the next few thousand years, plants reclaimed the exposed land, and animals followed in their stead. Larger animals such as woolly mammoths, musk ox, and caribou were readily able to endure the harsh climate, while the cold ocean waters teemed with mammals and fish alike. A land rich with so many resources—and relatively few predators—was ideal for human habitation, and the early Wabanaki peoples made the land that would later become known as Maine their home.

Of the four Wabanaki tribes, the Peskotomuhkati (Passamaquoddy) inhabited the eastern land. Directly next to Eastport is the tribe's primary summer settlement, Sipayik. Here, the tribe thrived on fishing the coastal waters, spearing giant pollock and smoking the oily fish for storage. Haddock, cod, and mackerel could easily be caught from the shore, while salmon and alewives ran freely down the nearby St. Croix River.

The Passamaquoddy tended to use islands like Eastport for storing their food through the winter to keep it safe from predators (including gray wolves, cougars, black bears, and coyotes). In the summertime, Eastport's ample shorelines—particularly during the 26-foot-deep low tides—provided exceptional opportunity for harvesting seafood and shellfish.

Porpoises and whales proved greater challenges, but skilled hunters rowing strong and fast birchbark canoes were able to bring them down, providing a feast for the village when they did.

The Passamaquoddy developed a complex language, in addition to a rich culture that included hundreds of dances and songs recounting the stories of legendary tribal heroes, such as Glooskap the Creator. They utilized animal hide and plant products to create durable garments, housing, traps, and baskets, some of which were dyed in an array of colors. Tools and weapons made from shells, stones, and copper (traded from inland tribes) enhanced their daily lives, while the land's remote location served as an effective deterrent to hostile newcomers for hundreds of generations.

A petroglyph of a Passamaquoddy person making a gesture of welcome (with hands raised and open palms facing the viewer) is among the more than 700 still visible at Machias Bay. Now owned by the Passamaquoddy tribe, the site represents the largest petroglyph repository in North America and is currently being carefully combed over by tribal cultural preservationists seeking to identify and catalog new findings. Estimated to be more than 3,000 years old, this petroglyph includes the detail of the person's hand being split in a V, making the sign for antlers that are sometimes worn by the tribe's spiritual dancers. This photograph is of a replica of the original petroglyph that viewed and touched, along with the others shown later in the book, at the ranger station of the St. Croix Island International Historic Site in Calais. (Author's collection.)

While no photographs exist of the way Eastport and its surroundings looked during the early era of the Passamaquoddy people, these images from the mid-20th century give an impression of the dramatic shoreline that sees evergreens rising directly upwards from craggy granite shores. Per Passamaquoddy legend, the larger island and the smaller islands near it were formed by Glooskap the Creator when he transformed a moose and the wolves chasing it into islands. The big island (seen near the top of the above photograph) was called Moose Island thereafter, while the Wolves Islands lie in the bay. Deer Island (New Brunswick) is on the right, with Indian Island protruding below it. Between the Moose, Deer, and Indian Islands, Old Sow churns the waters of the bay.

Eastport and the nearby lands were greatly affected by the Laurentide ice sheet during its long dominion. When the sheet pulled back, it left various land formations indicating the extent of its weight and power. Among the land formations still visible are eskers, resembling raised ridges or inverted riverbeds. Each one was created by an ice-walled tunnel at the bottom of the glacier that served to channel streaming water and accompanying gravel. The accumulated glacial debris remains today in the form of eskers located along the coastal inlands, to the delight of glacial geologists. Some esker ridges are as high as 40 feet, and the rocks they deposited are smoothed from tumbling down the glacial channel. For the early inhabitants of Maine, including the Passamaquoddy, eskers have served as convenient bridges across lakes, rivers, and inlets. (Courtesy of Joseph Kelley for the Maine Geological Survey.)

Over the past several thousand years, Maine's coasts have been steadily sinking in tandem with rising seawaters. As a result, more and more islands have been created as their natural isthmuses disappear. Located just off Eastport and accessible only during a 2.5-hour low tide window, Matthew's Island is one of Maine's 3,166 offshore islands. The 14-acre island has long served as a home for migratory sea birds of various kinds, and the Passamaquoddy would have harvested eggs there as a result. The nearby Carrying Place Cove, meanwhile, is so named for the access it granted Passamaquoddy to carry their canoes to Cobscook Bay's more open waters during low tide. Later, the island was routinely burned to help encourage the growth of blueberries. Today, native raspberries and blackberries can grow abundantly there in the early fall. (Courtesy of the St. Croix Historical Society.)

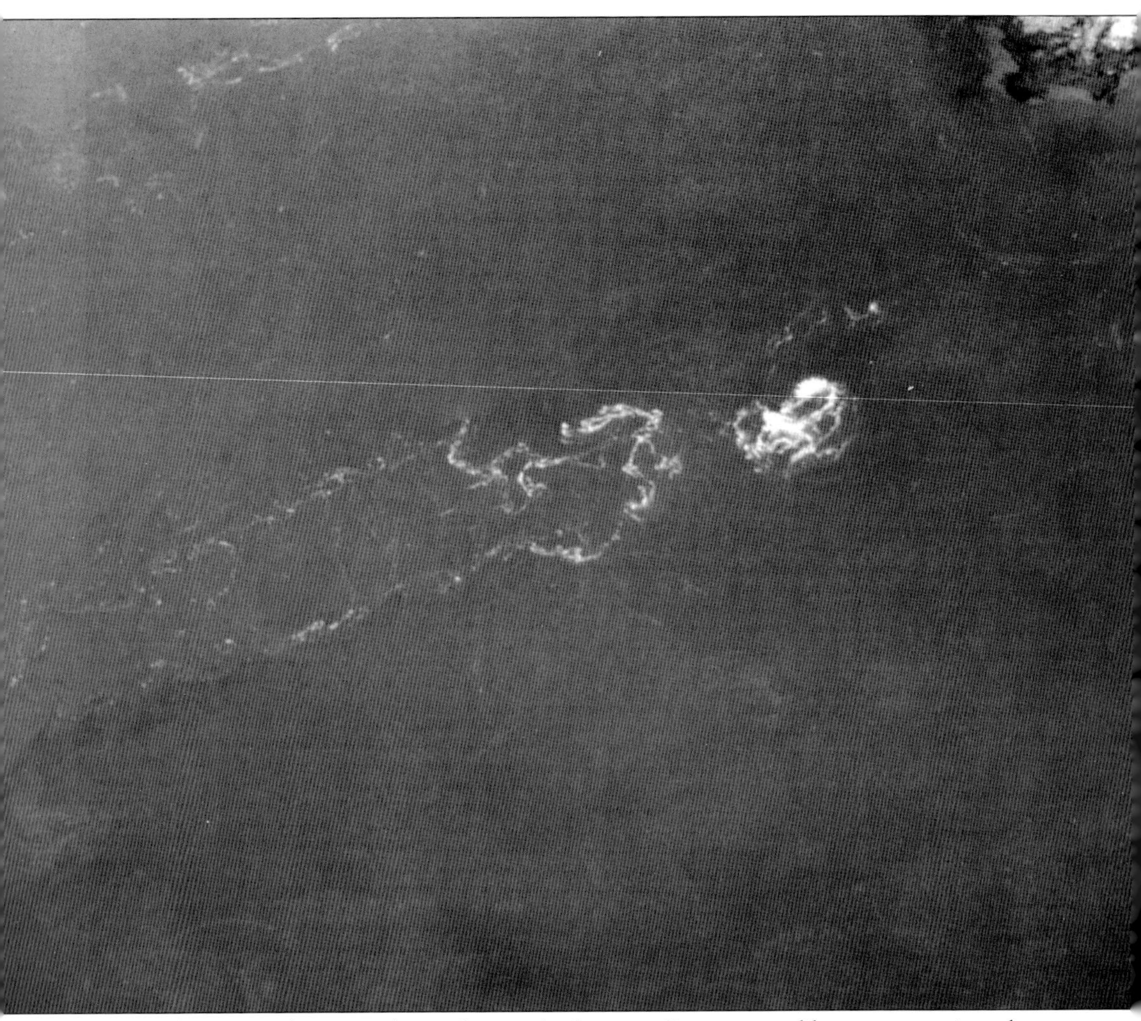

Just off the coast of Eastport loomed a steady threat to Passamaquoddy canoes passing close to its churning waters: Old Sow, the biggest whirlpool in the western hemisphere. While rarely composed as a single visible vortex, Old Sow can easily grab an unpowered vessel and spin it around until the tides change again. The strong currents and their constant duel with the flowing tides produce dozens of swirling eddies (called "piglets"). For unwary hunters traveling in canoes, Old Sow represented an unwelcome battle, as fatigue could easily result from attempting to escape its grasp. The early Passamaquoddy believed the whirlpool was made, like the islands around it, by the hand of Glooskap. After shrinking all the fish in the bay so they would eat less and be easier to catch, Glooskap met one who refused to be shrunk. "You want to be hoggish and eat everything? Now you will swallow the whole bay twice a day," he said to the fish, trapping it in place.

Lying on the shores of what would become known as Passamaquoddy Bay is one of its richest resources: beds of rockweed providing shelter and sustenance to many of the water's inhabitants. During high tide, the rockweed's bubble structures attempt to float to the surface as its holdfast keeps it rooted in place, resulting in a veritable life-supporting forest. When the tide goes out and the beaches become exposed, the rockweed collapses in on itself to create folds of moisture that make ideal habitat for shellfish, amphipods, eels, and various other organisms as they wait for the tide to flow back in. The intertidal zone is tasked with enduring a full range of conditions, from hot and extremely salty (caused by evaporating tidal pools in full sun) to cold, fresh water (falling from the sky or running from streams). As such, the animals that make the zone their home are among the hardiest in the world. Rockweed has the added function of filtering nutrients for the bay, creating layers of habitat from the ground up. (Courtesy of Robin Hadlock Seeley, Maine Island Photo.)

The cold waters around Eastport teem with phytoplankton, and as a result, whales of various sizes and species frequent the bay. Humpbacks like this one can leap fully out of the water, while the 70-foot-long fin whales are second in size only to the great blue whale. Minke and right whales round out the usual roster. The Passamaquoddy, rather than opting to fight these dangerous and intelligent giants in open waters, would lure them into shallow coves to reduce their maneuverability—and thus lower the chances of having their canoes flipped or crushed under the weight of a breaching behemoth. Once there, hunters would harpoon the whale from their canoes until it could persist no longer, at which point it would be towed to shore. When successful, whale hunts provided a substantial amount of food, blubber, bait, and crafting materials for the tribe. (Courtesy of Don Dunbar, Eastern Maine Images.)

Fish weirs, a common tool utilized by Native American tribes for several thousand years, were favored by the Passamaquoddy due to the high tides in the area. The weirs can be constructed by placing tall posts or tree trunks on the beach just above the lowest tide point and then stringing nets between the poles other than in one space left for the opening (facing the beach). When the tide comes in, fish fill the waters, some of them swimming back toward the open ocean and finding themselves ensnared in the netted walls or seemingly trapped in the small area. As the tide goes out, the Passamaquoddy could easily spear the loose fish and harvest the trapped ones from the nets. This method of fishing greatly supplemented the tribe's diet. Fish weirs are so effective that their use is continued today (with the postcard above based on an image from the early 20th century). (Courtesy of the St. Croix Historical Society.)

Wildlife is equally abundant on land in northeastern Maine. White-tailed deer were heavily hunted by the Passamaquoddy along with moose and caribou. The deer were prized for their meat as well as their skins, which were turned into clothing or used as insulation or bedding. The hides were dried and used as leather, and their antlers served as tools and weapons. The North American porcupine, meanwhile, was hunted in part for its quills, which could number up to 30,000 on a single animal. The quills were used as decorative elements in everything from clothing to basketry. Inland, the Penobscot—one of the Wabanaki tribes—would roast porcupines whole to prepare the fatty meat for feasting. Both of these photographs were taken by famed nature photographer William Lyman Underwood in 1896.

This replica of a petroglyph shows a spiritual woman of the Passamaquoddy tribe. Called *m'teowin*, the spiritual power of the Passamaquoddy is based in an understanding of interconnection and refers both to the power itself and to those who possess it. In this case, the m'teowin woman has called two wolves to her side, most likely for protection or to help her hunt. Some tribes were affiliated with particular animals (either due to relying on them as prey or to take on their aspects), while individual m'teowin could be bound to spirit animals that respond to their communication. *Malsum*, which means "wolf," was applied to Glooskap's evil brother in some Iroquois legends, but the Passamaquoddy did not view the wolf as a malevolent figure. In one legend, a wolf provides a traveler with the ability to make fires to help keep himself warm. (Author's collection.)

These historic portraits of two Passamaquoddy elders—Mary Selmore and husband Chief Sofiel (Sapiel) Selmore, as photographed by Charles E. Brown—encapsulate the tribe's enduring spirit and its culture. Women were typically the farmers of the Wabanaki tribes; they cared for the children when the men were hunting or gathering away from the home. They would often make the meals and prepare meats for smoking. Along with hunting game and foraging for food, the men would defend the tribe's lands—though the Passamaquoddy rarely participated in major warfare. Both men and women engaged in dancing, singing, artwork, and traditional medicine applications, and both could hold positions of power in guiding the tribe. Traditionally, only men were chief (though that has since changed). In these photographs, both genders are wearing tribal jewelry made from beadwork and hammered metals.

Shown outside of his temporary wigwam (or home), William Nicholas is pictured wearing traditional deerskin garments, an ornately embroidered headband with large feathers stuck into it, and fringed leggings that would have emphasized his movements while he was dancing. On his feet are moccasins, providing comfortable foot protection and a softened footfall ideal for stalking prey. The cone-shaped wigwam is made from a simple wooden frame with a cloth hung around it. Nicholas is surrounded by handwoven baskets, which would have traditionally been used to carry foraged materials, meats, fish, and a variety of other items but were later a primary source of income after colonial powers arrived and tribal lands were taken over. The banjo is not traditional, as it was developed by African American slaves, though it may have served to put Nicholas's potential customers more at ease. On his left is a rifle, which greatly increased the firepower of Passamaquoddy hunters and was among the more sought-after traded goods during earlier transactions.

This photograph of Chief Horace Nicholas and his family contains many traditional elements, albeit post-contact. The family wigwam is much larger; it is made from birchbark and may have skins lining its interior. The family members have incorporated traditional garb into their outfits, with the parents showing the clearest examples of how their ancestors would have appeared. Giant bones, antlers, and skins are presented for sale with a plethora of woven baskets resting in the center. Passamaquoddy children would typically participate in hunting and fishing, along with being assigned chores at home. During times of relaxation, they played with dolls or with toys made from sticks, stones, and balls. Passamaquoddy basket weaving is some of the finest in the nation, with this modern example by D. Campbell demonstrating the craftsmanship that goes into building the ash and sweetgrass creations.

Two

Growth of an Early American City

While the Passamaquoddy people may have had some limited interactions with Vikings and fishermen venturing across the Atlantic Ocean in the centuries prior, the majority of early European contact occurred in the 17th and 18th centuries. Once it did, the era of the indigenous people in North America changed over the course of a few generations to one dominated by European influences.

In accordance with the French quest to establish New France (and in particular Acadia), Eastport fell under the territory assigned by French kings to explorers and noblemen throughout the 1600s. It began in 1604 when Henry IV sent Pierre Dugua to the area. The group settled on St. Croix Island, located about 18 miles upriver from Eastport, and continued through to 1684, when Jean Sarreau de St. Aubin was granted Passamaquoddy Bay by Louis XIV. Sarreau had been living in Passamaquoddy, as Eastport was then known, since 1676.

For the most part, relations between the Passamaquoddy and the French were amicable. The French Jesuits treated the chiefs with respect, and the two groups exchanged goods and culture freely. Jesuit priests, traders, and hunters lived at Sipayik, and Catholicism quickly became the region's dominant faith.

In 1704, however, Eastport and its surrounding inhabited French and Passamaquoddy communities were plundered and dispersed by a British contingent sent from Boston under the leadership of Col. Benjamin Church as part of Queen Anne's War. Six hundred soldiers and two man-of-war ships (*Jersey*, with 48 guns, and *Gosport*, with 32) took all of the settlers they could find as prisoners.

When the War for Independence came, the Passamaquoddy threw their support behind the rebellious Americans to unseat British power in the region. After the Revolution, however, the border was in dispute, with both sides claiming ownership of the area. In 1814, the British seized Eastport for four years before returning it to the former colony. From then onward, the close proximity to Britain (and later Canada) meant that Eastport was a haven for smugglers and those attracted by the wild coastal frontier.

For the next half-century, Eastport would see its population boom as the town looked to a horizon bright with the promise of abundant natural resources.

The first depiction of a European ship comes in the form of one of the last petroglyphs recorded at Machias Bay. An accurately scaled ship with a mast and a sail is shown at left, and a Christian cross is pictured at right. The oral history of the Wabanaki recounts their amazement at the size of the ships, which would have been primarily French during the first century of significant contact. Jesuit missionaries spread the Catholic faith throughout tribal lands, and the tribe was receptive to the symbolism and ritual the religion offered. While the Passamaquoddy were spared some of the loss of life experienced by southern tribes in English lands, the Great Dying of 1616–1619 (caused by foreign diseases) was still devastating for the tribe. In 1615, the Passamaquoddy tribe was estimated to have about 1,400 warriors living at Sipayik; by 1820, a total of 400 Passamaquoddy (men, women, and children) remained. The age of sail was officially underway. (Courtesy of John Jackson.)

In 1772, Eastport was formally established by James Cochrane and a cluster of families from Essex County, Massachusetts. They built houses, docks, and buildings for curing fish, and the village grew at the east end of the island. Initially, the only method of travel was by water, though High Street was built in 1799 and horses arrived in 1804. Upon seeing one for the first time, a young boy remarked to his mother, "There goes a man sitting on a cow that ain't got any horns," per William Henry Kilby in *Eastport and Passamaquoddy*. Capt. John Shackford, a hero of the Revolutionary War who accompanied Benedict Arnold on the ill-fated march to Quebec, came in 1783, followed soon after by Arnold himself (after his act of treason at West Point). Arnold went on to conduct a prosperous smuggling trade in the area.

The early settlers found Eastport highly temperamental in its climate, with the proximity to the ocean causing rapid temperature fluctuations. Fog was a nearly constant presence in the summer months, and sea smoke floated freely from the ocean's surface throughout the winter. Hay was harvested in the marshes, and some settlers grew potatoes in the acidic soils. This lithograph by Stow Wengenroth encapsulates coastal Maine living in the era.

In 1802, the weekly mail operations were coordinated by Col. Oliver Shead, the island's first postmaster, horse owner, and builder of a two-story house (still standing at 130 Water Street). The need to get to the mainland easily became apparent, and in 1820, the first wooden toll bridge was built. Spanning between 1,200 and 1,400 feet long, it replaced Samuel Tuttle's ferry service in functionality. In 1832, a second bridge was built, this time running through the Passamaquoddy village of Sipayik. (Courtesy of the St. Croix Historical Society.)

Recognizing the vulnerability of Eastport due to Britain's close presence, and with the northern border still disputed, the young American nation raised Fort Sullivan in 1808 on the top of the island's central hill (officers' quarters shown above). When war was declared with Britain in 1812, a third of the island's population left in anticipation of conflict. The fort saw its first action in 1813 when a captured British ship, the *Eliza Ann*, was brought into the port for safekeeping by the privateer *Timothy Pickering*. The vessels were pursued by the British sloop HMS *Martin*, the commander of which demanded the return of the *Eliza Ann*. When the fort's commanding officer refused, the *Martin* opened fire on Eastport; the troops of Fort Sullivan immediately retaliated with artillery and drove the British ship away. The following July, a different story would unfold, as a 10-vessel British fleet led by HMS *Ramillies* (with 74 guns) demanded the surrender of the island as a whole—and, within five minutes, Eastport obliged.

The British held Eastport from 1814 to 1818, with the consensus being that the occupation was generally a civil affair. The British claimed Fort Sullivan and built a powder house on-site, the structure of which would endure another century and a half as a favorite place for children to explore. The British expanded Fort Sullivan, though many of the larger additions were later removed. On June 30, 1818, the border was settled. Eastport was handed back to American troops, who fired a 21-gun salute as "Yankee Doodle Dandy" was played and the American flag was raised over the fort. The next year, the *Eastport Sentinel* printing offices were established, granting the fourth estate fertile ground to grow from. Eastport's population rose to 1,937 in 1820, and within a decade, the state counted 209 homes, 72 barns, and 34 stores or shops, per William Henry Kilby. In 1828, an extensive saltworks operation was established on the southern part of the island, and hundreds of men were soon employed in harvesting more than 1,000 bushels of salt a day.

In the 1790s, interest in holding religious meetings was sounded. The Baptists were the first to build a permanent church structure. Dedicated on December 1, 1819, the North Baptist Church was built at the top of Washington Street, with the first sermon presided over by Elder Samuel Rand. Inside, William Henry Kilby records how the building was heated in an unusual fashion, with the stoves hung in midair and attached to the supporting columns. Col. John Shackford was among those signing for the church's incorporation, as was his son William, who was an adventurous seafaring captain by this time. After being captured in 1812 while transporting rice and flour to Spain, William was returned to Eastport on one of the last boats allowed to come from Europe to America after the declaration of war in 1812. (Above, courtesy of Special Collections, Raymond H. Fogler Library.)

The Catholics established their own society in 1826 under the guidance of Rev. Charles French, who organized a visit to the island by Bishop Joseph Fenwick of the diocese of Boston the following year. A church was quickly planned, and the cornerstone was laid in 1828. By the following year, St. Joseph's Catholic Church was completed at the corner of Boynton and Chapel Streets. After 50 years, the original church proved too small, and it was moved and converted into a dwelling while a new edifice was constructed in 1873. By 1887, the church added 1,850 square feet of space in addition to lifting the entire structure by two feet and giving it a brick and stone foundation. The ornate interior offers a wealth of Roman Catholic iconography in the form of stained glass, paintings, and statuary. (Left, courtesy of Special Collections, Raymond H. Fogler Library.)

In 1828, the First Evangelical Congregational Church and Society of Eastport was organized after a decade of informal meetings. The following year, the Central Congregational Church was built on Middle Street. Designed by congregation member and architect Daniel Low, the church is considered unusual for the time period in that it exemplifies Federal-style architecture, a rare trait for frontier communities. Rev. Wakefield Gale conducted the first sermon with the first deacons being Ezekiel Prince and Libbeus Bailey. The Congregational church was the first in Eastport to have its vestry under the same roof. The town relied on this church to provide accurate time and supplied funds for the purchase of its clock. The original steeple blew over in the Great Saxby Gale of 1869, but a similar one was set in its place.

During the British occupation, Anglican services were provided to those seeking them by the Church of England. Afterward, the American Episcopal Church led services in the other meetinghouses of the town until 1857, when Rev. William Stone Chadwell started holding regular services for a growing number of attendees. That same year, the Christ Church congregation was established, and construction on a church on Key Street began a few months later at a cost of just over $3,000. Once the church was completed, the townsfolk took to calling the attached parish hall "The Institute," a colloquial name that continues to this day as a wide variety of community services and meetings are held in its quarters. Other churches in the area include the First Congregational (Unitarian) Church (dedicated in 1820), Washington Street Baptist Church (dedicated 1837), and the Methodist Episcopal church (organized in 1838).

Eastport's prosperous families had numerous sprawling houses built over the island's lower streets, including merchant J.K. Norwood, who commissioned this house at 15 Key Street. It was built in 1824 in the Greek Revival style, complete with Doric columns, and the Norwoods enjoyed it for more than three decades. In 1866, the house was inhabited by the Whelpley family; in that year, the famed Civil War hero Maj. Gen. George Meade—who commanded the Union army at Gettysburg—stayed with the Whelpleys to keep watch over Eastport during the Fenian Rebellion. While Eastport did not come under threat, the general reportedly did, as Mrs. Whelpley was notedly unimpressed with his habit of carelessly spitting chewing tobacco around her home. While staying with the Whelpleys, General Meade used the portion of the first story facing the street for his quarters.

In 1869, another of Eastport's iconic houses was built, this time in the Gothic Revival style at the behest of Dr. Luther Babb and his wife, Eliza. The Babb house was constructed out of granite most likely harvested from Red Beach, a small town up the St. Croix River toward Calais, and transported by barge. Red Beach would go on to claim one of the most advanced granite polishing operations in the world a few decades later. Luther and Eliza moved to Eastport in 1861, with Luther having a goal of building a medical practice in the East after spending years in the West. Eliza earned her own doctoral medical degree in 1871 from the Women's Medical College of Philadelphia. Their daughter Cora became a doctor as well, and their other daughter, Grace, attended the Philadelphia College of Pharmacy and was the second woman to graduate. The family played a significant role in exemplifying educated women in the community.

As Eastport grew, it was divided into three school districts, though these were disproportionate in terms of population. By 1834, Jonathan Weston recorded 70 scholars between the ages of 4 and 21 in the town's north district, 84 in the middle district, and a whopping 970 in the south district where most of the residences were located. Upon entering their teenage years, students attended the Old South School on High Street in the center of town. After a fire destroyed the Old South School in 1846, Boynton High School was built to provide Eastport's teenage students with two modern floors of learning. The building's Italianate design was created by notable Boston architect Gridley James Fox Bryant, who was commissioned for two other structures on the island (neither of which survive). It functioned as a school for more than 70 years before being claimed by the community for other organizational purposes, including as the city hall for many years.

One of the more charming early Eastport narratives is Eliza Allen's. Born in 1826 to what she described as a "respectable and wealthy" family, Eliza Allen fell madly in love with William Billings, a lower-class boy. Fleeing Eliza's outraged parents, William left to fight in the Mexican-American War—and Eliza disguised herself as a man and enlisted to find him. After doing so, the pair returned and were wed in Eastport with her parents' blessings. (Courtesy of Special Collections, Raymond H. Fogler Library.)

Channeling Maine's strong abolitionist beliefs, Eastport contributed 403 soldiers to the Union army during the Civil War. Unfortunately, many did not come back. As part of the 1st Maine Heavy Artillery Regiment, for example, many sons of Eastport were involved in a fateful charge during the 1864 Siege of Petersburg, which saw the greatest loss in a Union regiment in a single day of combat during the Civil War.

In 1868, one of the more important cultural buildings of the era was completed: the Memorial Hall & Opera House on Boynton Street. Traveling performers from across the young country would come to entertain the citizens of Eastport and those visiting its harbor. The famed minstrels Culhane, Chace & Weston were among those who delighted with their vaudevillian routines, while Punch and Judy shows (dating back to the 1660s) drew raucous laughter even as they mockingly portrayed interspousal violence. Others came to see talented local and regional athletes competing at basketball, which fast became the beloved sport of eastern Maine. The Memorial Hall was the largest public space available in the town, and it served virtually every function as a result—from the firemen's ball to the Catholic Christmas Mass—until it burned down on January 7, 1913. (Courtesy of the St. Croix Historical Society.)

Another major building completed in 1868 was the Passamaquoddy Hotel. Built on the corner of Key and Water Streets, a 24-foot-wide road constructed in 1803 that was (and is) the primary route for commercial traffic through Eastport's downtown, the Passamaquoddy was one of the largest hotels in the half-century-old state. The town's population had grown to just over 3,700 by this time period, as fishermen, sailors, merchants, and laborers found steady employment in the growing local industries. An infrastructure of 100 warehouses and stores filled the downtown, providing ample opportunity for trading and shopping to those visiting it regardless of their persuasions. The back-to-back completion of the Passamaquoddy Hotel and the Memorial Hall made a trip to Eastport all the more appealing for the next few decades, and travelers from up the St. Croix River, down the coast, and points inland made the adventure frequently via steamships and carriages alike.

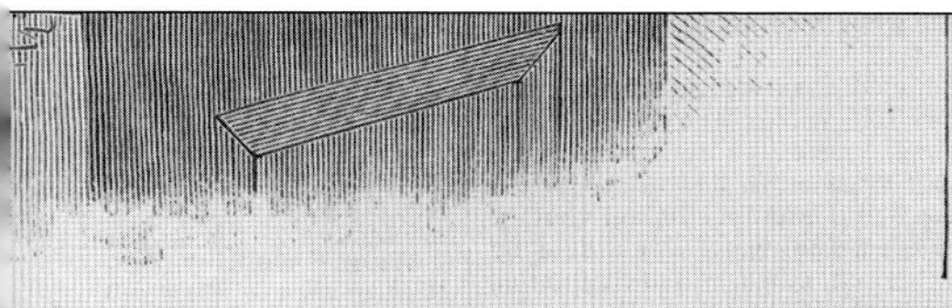

OR-GENERAL JOHN SEDGWICK BY THE SIXTH CORPS.
OTOGRAPHED BY ROCKWOOD.]

Eastport Island, where Passamaquoddy Bay is connected with Lake Utopia by a marsh a quarter of a mile long. Being attacked by musketry, it struck for the marsh, and probably for the lake, which was undoubtedly its home, and before being rendered incapable of locomotion, it had worked its way with its fins and legs a number of rods. The report of its presence at once spread to the town, attracting a large number to the spot to aid in its destruction. It received some seventy musket balls, and although attacked in the forenoon, it exhibited signs of life the following day.

"Thus the northeastern point of our State, with the assistance of New Brunswick, has the honor of producing the nearest approach to a veritable sea-serpent, which is destined to make a popular sensation wherever exhibited. It is to be at Portland during the forthcoming State Fair, and is thence bound for Boston, New York, and other principal cities."

THE WONDERFUL FISH, CAUGHT NEAR EASTPORT, MAINE, AUG. 3, 1868.—[DRAWN BY CHARLES A. BARRY.]

With its close connection to the wild depths of the sea, Eastport was well known as a place of mystery. While locals were familiar with a wide variety of creatures—from great white sharks to fin whales—they would occasionally come face-to-face with animals they could not readily identify. Such was the case in August 1868, when the *Eastport Sentinel* reported that four fishermen had pulled in an unusual large fish measuring 30 feet long and over five feet wide with five rows of gills and a blunt nose. It soon became known as the great shark-dog fish, and, once stuffed and mounted at the Boston Zoological Institute, it was dubbed the much more colorful great Utopia Lake sea serpent (due to an earlier sighting at Lake Utopia, New Brunswick, where it presumably walked from). The excitement surrounding the creature was significant, with one offer to purchase its hide amounting to almost $50,000 in 2021 currency. Later reports suggest the fish was a basking shark, which can grow up to 40 feet in length. (Courtesy of the St. Croix Historical Society.)

On the evening of October 4, 1869, an ill wind blew, and Eastport was directly in its path. The Great Saxby Gale combined with the extremely high perigean spring tide (predicted by Lt. Stephen Martin Saxby, who would go on to lend the hurricane his name) to create a tidal wave that swept through Eastport's downtown. In the morning, the townsfolk gathered to survey the aftermath, including a number of damaged buildings and waterlogged trade goods. Traveling between Eastport and Calais, one man counted 90 houses destroyed, and the storm was declared the worst to have hit the area up to that point in recorded history. Altogether, from New York to New Brunswick, Canada, more than 37 people died as a result of the combined raised waters, steady rain, and high winds.

Just inland over the bridge in Sipayik, the Passamaquoddy people struggled to retain a sense of tribal identity as the heavy hand of the state continued to be felt. By 1819, the St. Croix River—representing the ancestral river of the Passamaquoddy people—had 47 mills along its shores, and it was abundantly clear that the indigenous peoples were no longer sovereign in their own land. In 1833, the tribe's Wabanaki brethren, the Penobscots, had the last of their townships (representing 95 percent of their existing territory) taken from them by the state despite a treaty with Massachusetts from 1818 declaring otherwise. At Sipayik, a disagreement over leadership split the Passamaquoddy tribe into two, and a contingent left to permanently reside at the tribe's winter home in Motahkomikuk. To survive the loss of traditional lands used for hunting, fishing, and dwelling, the Passamaquoddy turned to crafting baskets for industrial and ornamental purposes or assimilated into employment in the waterfront fisheries.

Downtown Eastport was bustling with activity in the 1860s, regardless of the time of year. The predominantly wooden, multistory buildings lining Water Street provided a handsome skyline that indicated Eastport's prestigious place in the hierarchy of flourishing New England port towns. Residents and visitors were able to purchase any number of goods or services in Eastport, including many imported wares (as about half of all incoming vessels were from foreign ports). This made it a favored destination for both those looking to establish a quality lifestyle and country farmers seeking supplies to restock their stores. Restaurants, hotels, clothiers, and dentists are among the businesses visible in this photograph. In the winter, while the ocean effect mediated the amount of snow received, the small physical space allotted to the downtown and its adjoining streets meant that snow could easily pile up. As this photograph makes clear, however, modest snowfalls did little to deter the pace of commerce Eastport had become accustomed to.

Three

Of Ships, Sardines, and Shanty Towns

By the late 19th century, Eastport had become a flourishing—if remote—port town, due in part to its direct connections to Boston by boat. All of the newest inventions and modern goods available in the "Athens of America" to the south were thus available in the modest island town, making Eastport the best shopping center for clothes, shoes, and tools east of Bangor.

While doing its steady mercantile business, Eastport suddenly became the hub of an even bigger industry: North American sardine canning, which got its start in Eastport in 1876. Massive amounts of manpower—in addition to work by both women and children—became devoted to catching, cleaning, and canning the giant schools of small herring that swam around Passamaquoddy Bay.

The launch of the sardine industry further propelled the activity of Eastport's port, which, as of 1833, was second only to New York City. The island town had gained a name for itself, and news from it occasionally reached papers across the nation.

Eastport's financial successes were not equally shared amongst its citizenry, and the living conditions between the wealthy and the working class were starkly contrasted by the island's small physical size. To keep the sardine industry going, hundreds of Eastport residents and transient families from outlying communities such as Perry and Calais joined families from Back Bay, Deer Island, and Campobello in Canada to put in long days for low wages, particularly when the herring were running in late summer.

For decades, the canneries were synonymous with Eastport, as was the accompanying smell of fish waste products. Other industries cropped up in this time frame, including a shoe factory and what would eventually become a renowned mustard mill.

Eastport was occasionally faced with significant challenges, including three major fires in the 19th century, the last of which led to the downtown being reconstructed mostly from brick.

The overall prosperity of the island was enough for Eastport to be incorporated as a city on March 18, 1893, and it adopted a city council and a city manager to govern it.

In 1876, Eastport celebrated its first 100 years of official existence as a town. Just over a decade past the Civil War, the messaging was fiercely Unionist with a banner running across Water Street reading "Liberty and Union / One and Inseparable." By 1880, Eastport counted 4,006 residents with estates valued at $888,892, owing in part to the port's flourishing cod and pollock fisheries.

In many ways, the ocean-hardened face of the rugged fisherman defined the next era of Eastport's existence. The same year it celebrated its centennial, Eastport saw the opening of the first sardine cannery in North America, with the Wolff & Reesing Cannery starting operations on February 9. Previously, sardines were imported from Europe, but the Franco-Prussian War disrupted the shipments and prompted the need for a new source.

The herring fishing fleet of Eastport was vast, consisting mainly of small vessels known as "mosquitos" that ran daily in and out of the harbor during the summer and fall. Their outgoing destination was the weirs surrounding the island, which were typically full of sardines; once there, the fishermen would scoop up the small fish in nets until hitting capacity (at 5,000 or more fish each). When returning to the docks, the small vessels would sometimes require towing if the tide, current, and winds did not align favorably. Some of the mosquito fleet were owned by Canadian fishermen, who were happy to share in the profit by bringing their fish to the awaiting canneries. Among the steam-powered vessels that helped with towing operations large and small were the *Luce Bros*, *Henry F. Eaton*, *R.J. Killick*, and *Mary Arnold*.

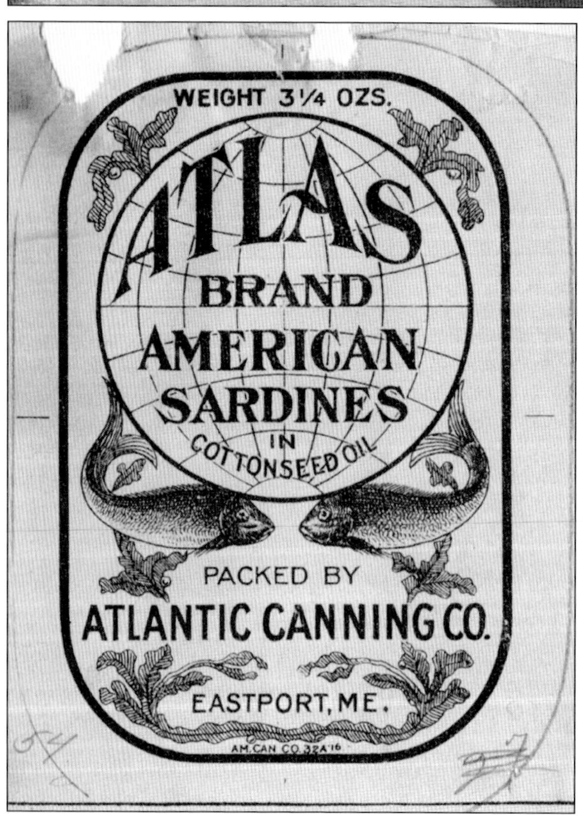

At its peak, Eastport had 18 canneries extending far out into the bay. The large, multistory buildings hummed with activity from May 15 to December 1 as hundreds of workers toiled at various stations. The smallest herring, measuring between five and seven inches, were cut and packaged as sardines, while larger fish were cut into chunks and sold as steaks or filleted, smoked, and packaged as kippers. The can labels evoked the aesthetics of the era, lending bold Victorian and Art Deco stylings to each brand. The Atlantic Canning Company was among the first; it produced up to 70,000 cases a year until closing its doors in 1923. Individual factories sold out to larger brands such as American Sardine and Seacoast Canning, though some—such as L.D. Clark and Sons and M.C. Holmes Canning Co.—preferred their independence.

Children in their single digits were among the workers who spent their days (and sometimes nights) in the sardine factories, some of whom were documented by photographer Lewis Hines. The knives the children used matched those of the adults, and badly injured or severed fingers were common. The workers typically stood in ankle-high fish waste throughout their shift. These boys at the Seacoast Canning Factory No. 4 were between 10 and 12 years old and earned 5¢ a can, with some younger children filling only five boxes or so a day. Skilled teenagers could earn $2 or more a day but only by working late nights. Women and men comprised the remainder of the workforce, as shown in a scene from the B.H. Wilson cannery. The number of laborers grew to 1,200 at peak. (Above, courtesy of the Library of Congress.)

Along with its fisheries, Eastport maintained various other industries. Lumber was—and still is—one of its primary exports (as this ship in the center of Prince Cove demonstrates), going back to the English and the harvesting of 200-foot-tall white pines for ship masts. As Eastport matured, it developed a robust shipbuilding and drydocking reputation, with the clippers *Grey Feather* and *Crystal Palace* among those built on the island. Third-generation shipbuilder Caleb Stetson Huston was the most noted producer, owning four shipyards and building over 100 vessels until his death in 1887. Shoemaking was a popular industry that offered an alternative to the fishy business of the waterfront. The Holmes Shoe Factory, located toward the west end of the island, employed hundreds of men and women as they produced thousands of shoes for local sale and export.

Around 1885, Bank Square on Water Street was a thrumming, developed district that offered a slew of services to the discerning customer. On the right near the foreground is the ornate Masonic temple, made from wood, followed by the facade of the Frontier Bank. Up the hill is the Passamaquoddy Hotel, and at center is the original Eastport Savings Bank building.

Toward the opposite end of Water Street, the Sentinel Building stood prominently on the left. On the top it housed the *Eastport Sentinel*, established in 1819; on the bottom, the Wadsworths operated a hardware store that would ultimately extend beyond six generations (through today). Founder Samuel B. Wadsworth was the son of the famous Revolutionary War general Peleg Wadsworth and uncle to Henry Wadsworth Longfellow.

On October 14, 1886, a new level of calamity brought Eastport to its foundations and placed the small port in the national eye. A blaze caught at G.W. Capen's sardine factory along the waterfront and proceeded to consume the wooden buildings of the downtown. The telegraph lines were burned by the fire, and no dispatches could come out of the port, rendering its suffering silent yet luminous. In Calais, 30 miles away, townsfolk watched in sad horror as the sky glowed and

the fire raged unabated until it ran out of fuel. Altogether, the Great Fire of 1886 destroyed 130 stores, offices, and businesses; 62 houses; 17 wharves; 8 factories; 5 boardinghouses; and 2 hotels. While the ruins made for good grazing for local deer, it came at a high price: the *New York Times* estimated the damages to be $230,000, equivalent to $6.5 million in 2021.

Eastport was no stranger to fires, as the downtown had been previously decimated by two major blazes, one in 1839 and another in 1864. City firefighters were still vastly unprepared for the Great Fire of 1886, relying on two hand-pumped engines that were hard-pressed to keep up with the rate of combustion. To make matters worse, the fire sparked at low tide, and the water reservoir ran dry all too quickly. After the Great Fire of 1886, the fire department purchased new equipment for combating aggressive blazes. For their part, building owners rallied and replaced 26 buildings within the next year. Fortunately for the city, architect Henry Black—responsible for designing much of Woodstock, New Brunswick, after its fire a few years prior—had just moved to Eastport, and he had a hand in designing nearly two dozen downtown structures during the rebuilding.

Toward residential Water Street, the Great Fire of 1886 destroyed dozens of homes. Architect Henry Black once again stepped in, designing the house on the left (above) and several others. Serving as homes for the island's prosperous residents, many sported ells, barns, and outbuildings to shelter tools, livestock, and carriages. Their size and aesthetics stood in stark contrast to the shanties and small frame houses in the sardine camps a few blocks away. The house seen below belonged to the Goodell family; the father worked in a lumber mill nearby while Clarence (8), George (9), Lottie (12), and Violet (15) worked at the sardine canneries. The numerous sardine camps were often overpopulated as families from neighboring communities flooded in to find work, and diseases and viruses found easy purchase there in those who dwelled in uninsulated homes with no running water. (Below, courtesy of the Library of Congress.)

Inside the more lavish houses, the wealthier class held social gatherings and played fashionable Victorian games in parlors decorated with the finest goods from around the world. Patriotic decor was common, particularly around the holidays, and portraits of George Washington—highly venerated as the "Father of the Nation"—adorned walls in home after home in a display of adoration and loyalty. Spiritualism was gaining popularity, and seances and divination techniques such as palm reading were increasingly common. During smaller parties, Victorians played charades or indulged in chess, backgammon, and a growing list of other board games. Meals were augmented by cooking stoves (patented in 1834), hot water heaters, gas lighting, and sinks with hand-operated pumps, all of which would have been the realm of the servants in wealthy homes.

This Queen Anne–style home on Key Street is among Eastport's unique houses. Built in 1894, the house was commissioned by "Captain" George W. Capen. A captain of industry rather than of combat (though he served as a lieutenant in the Civil War), Capen owned a plant located at Market Wharf (at the bottom of Boynton Street) that was dedicated to decorating tin sardine cans. His was the sardine factory that had ignited to begin the Great Fire of 1886.

EASTPORT SAVINGS BANK BUILDING

The Eastport Savings Bank was rebuilt in brick on Bank Square following a design by architect Henry Black in the Italianate style. The building has been used for the Eastport Bottling Co., apartments, the police station, and city rooms; today, it is maintained by the Tides Institute & Museum of Art. At the beginning of 1880, the bank held approximately $153,780 in deposits; by 1890, it swelled to $348,648.

Following the Great Fire of 1886, the International Steamship Company built this large depot to encourage travel to connecting ports up and down the East Coast. Incoming tourists could take the ferry for a day trip to various towns around the bay, including Campobello, Grand Manan, St. Andrews, and Lubec, or to the towns along the St. Croix River. Passenger trains penetrated inland, extending the possibilities of locomotion for the Victorian traveler to new realms and making the resort towns of the coast more accessible than ever. The steamship *State of Maine* was built for the International Steamship Company by the New England Shipbuilding Company of Bath, and it served the region between Boston and St. John, New Brunswick, well until being decommissioned in 1924. Round-trip tickets to Portland were approximately $4 each and represented nearly 400 miles worth of travel.

Eastport's patriotism has long culminated in the largest Fourth of July festival in the state. Declaring the surrounding days to be Old Home Week, it coordinates a major parade and a slew of activities for all ages. The parades held after the Great Fire of 1886 were particularly joyous as townsfolk celebrated the restoration of their city in the face of widespread loss.

Constructed in 1887, again from the mind of Henry Black, the Beckett Building on Water Street is an Italianate in brick that housed Scotsman A.W. Beckett's confectionery store. The Beckett family had made a name for itself in Calais and New Brunswick as candy makers, and the legacy continued in the region through 1978. The second floor served as the lodge for the Independent Order of Odd Fellows.

Stores throughout the downtown enjoyed individual degrees of prosperity based on the type of goods they offered and their proximity to common routes. One clothing business, which L. Holitser operated since being established in 1878, was among the most competitive in the island town. The building's three floors were all dedicated to clothing manufacturing and custom work, and some 20 employees labored there. A contemporary advertisement promises "all the latest novelties" and a full line of "staple goods." Other stores were family-run affairs that offered a variety of foods (including fresh fruits and products made from local blueberries), barrels of necessities such as molasses and potatoes, equipment and supplies for the farm and the home, and canned foods from around the world.

Among the businesses that got their start at the beginning of the 20th century is Raye's mustard mill. Founded in 1900 by 20-year-old J. Wesley Raye, the business began in the family smokehouse when Raye recognized an opportunity in the vast number of sardine cans leaving the port every month. Developing his own line of mustard, Raye established a stone mill at the building's current location at the top of Washington Street. Today, the mill is run by the fourth generation of mustard-making Rayes and is the last still-operating stone-ground mill in North America. It serves as a working museum and ships Raye's Mustard across the world. (Right, courtesy of Karen Raye.)

After the Great Fire of 1886 decimated the famed Passamaquoddy Hotel, a new building lay claim to its location on Water Street in 1893. Named after donor Frank Peavey, the Peavey Memorial Library was built on the condition that it contain at least 5,000 books; as a result, townsfolk across the island donated from their own collections to get construction underway.

Following the loss of Eastport's customs office in the Great Fire of 1886, a new customs building, complete with a post office, was constructed using granite from Blue Hill. Completed in 1892, the building was designed to withstand fire while comfortably handling the port's customs receipts of over $100,000 a year. At the time, 14 sardine factories lay within a half-mile circle of the building.

The corner of Washington and Water Streets (with the customs building at right) represented the primary point of entry to downtown Eastport when arriving from inland, and nary a day passed without significant foot and carriage traffic. With close to 5,000 residents by this time, Eastport saw all manner of folk plying its streets during the day.

A few blocks away from Water Street heading inland, the neighborhoods were significantly quieter. This scene looks up Boynton Street to take in the large houses, wooden sidewalks, and relatively narrow streets that served to accommodate carriages but later became strained by modern vehicles. Boynton Street is notable in part for the proliferation of hip-roofed houses it hosts.

At the close of the 1800s, Eastport was a popular destination for cyclists and fishing enthusiasts alike. With the loss of the Passamaquoddy Hotel in the Great Fire of 1886, Eastport found replacements for lodging in the form of hotels such as the Riverside (with 20 sleeping rooms and a dining area that could accommodate 35 people), located across from the library, and the Quoddy on Washington Street, a spacious hotel with 65 rooms that could be rented for $1.50 a night (though the Quoddy was ill-fated and burned in 1904). Farther up Washington Street, Mabee's Lodging House was a former sea captain's home with 16 rooms that were nearly always full. Travelers could come in from the nearby Eastern Steamship wharf or via the Washington County Railroad depot (opened July 15, 1898) to spend a night or two in the port city.

As time passed, the need for an organized cemetery became evident, and so Hillside Cemetery was developed in 1819. The first to be buried in the cemetery may have been a man by the name of Haycock in 1820; others believe that British soldier Lt. Thomas Raymond, transplanted there in 1819, should be considered the first to be entombed.

Over the bridge in Sipayik, the Passamaquoddy reservation maintained its own cemetery; on the left, St. Ann's Church dominates the skyline. Pleasant Point, as English speakers called Sipayik, had only 252 people recorded for the 1900 census. The remaining families tried to hold on to their tribal identity while their children were forcibly sent to residential schools like Carslisle under Capt. Richard H. Pratt's "kill the Indian, save the man" philosophy.

With most of Eastport clustered on the east end of the island, residents who lived inland had farther to travel to get to necessities. For the children of the outlying areas, that meant taking a bus carriage to school each day along the toll bridge. The island city itself was always a source of entertainment and engaging antics, as evidenced by the arrival of the occasional wayward moose. Moose are excellent swimmers and commonly travel between the islands or swim to escape flies (though male moose are sometimes too encumbered by their antlers and end up drowning in the attempt). This female moose appears to have swum into the shore next to a sardine factory, much to the amusement of the men who found her.

Four

The Early 20th Century on a Coastal Frontier

At the beginning of the 20th century, Eastport was at the height of its station as an industrious port offering a bounty of worldly goods to travelers from all over the nation. Both the ongoing bounty of sardines (and the associated effort to pack and transport them across the globe) and a crop of new industries seemed to promise a bright future for the island city. For all intents and purposes, Eastport's gilded age was a prosperous one indeed.

During the census count of 1900, the island reached its peak population at 5,311 residents. This number is beguiling, however, as many more transient residents working in the canneries and the factories lived in the city part-time during the seasonal work periods. Beyond that, visitors and returning residents would fill the streets in the summer to spend time with relatives and do their weekly shopping errands. It was not uncommon to see upwards of 10,000 people in downtown Eastport, particularly during the Fourth of July (Old Home Week) when the summer festivities were at their peak.

The first half of the 20th century brought new and unexpected challenges for the country as a whole. While the remote nature of Eastport limited the impact of World War I, the Spanish flu, the Great Depression, and World War II to some extent, the ripples of each event were clearly felt, and Eastport joined most municipalities in Maine in declaring bankruptcy. After the Second World War, the recognition of the waning herring stock would seemingly write the fate of Eastport's prosperity in the sand.

But Eastport has always been more than a fishing community. Along with presidents, engineers, artists, and entrepreneurs, the visionary residents of the island town have continuously managed to find potential in the unique city, propelling it forward from one generation to the next.

The close of the 19th century symbolized the ending of a dramatic and highly transitional time period in Eastport. In the past 100 years, the population had swelled from 562 to 10 times that number. With the introduction of electricity to Maine (January 1880 in Piscataquis County), residents were no longer exclusively using candles and whale oil to light their homes; they now had access to electricity via wires strung throughout the city. Steamships offered new levels of transportation not only for the residents of the city but also those located far beyond who would come to visit the famed eastern port and partake in its exceptional fishing and shopping opportunities. For those traveling by land, riding on horseback and via carriages remained dominant for decades, and the bridges connecting Eastport to the mainland were the sole points of access. Beyond the interior bones of this toll bridge, papered with notices of events past and present and purveyors of sundry goods, Eastport's light proved alluring time and time again.

By 1916, Eastport's image as a jewel by the sea was complete. The devastation wrought by the Great Fire of 1886 thirty years earlier was entirely healed, renewing the city to new heights. The customs house—built from hewn granite cleft from the earth itself—dominated the center of downtown in an homage to the stalwart forces of industry and trade. Water Street, extending to the north and south along the shoreline (in the background of this photograph), was now lined with handsome brick buildings designed by the likes of Henry Black. On the global stage, World War I had been declared two years earlier, and, while it would be a year before the United States entered the fray, the nation's military forces were in preparation mode. Domestically, social conditions continued to improve, with Maine establishing a maximum 58-hour workweek for women and children and children under 16 now protected with child labor laws. (Courtesy of the St. Croix Historical Society.)

Key to the developing bustle of Eastport, both in terms of shipping passengers and freight, was the train station. The Washington County Railroad opened a line to Eastport in 1898, and Maine Central Railroad purchased it in 1917. The train station was located approximately between today's IGA and First Bank on Washington Street.

The women of the early 20th century made strides toward equality in part by emphasizing an egalitarian interpretation of Christianity. Bicycles represented freedom from the confines of the home and were adopted by suffragettes as a means of being active in society and politics. In 1919, as a direct result of the industrial responsibilities women took on during the Spanish flu, women gained the right to vote.

New to the streets of Eastport were horseless carriages capable of traveling long distances without requiring rest. The Model T arrived in Maine just after the turn of the 20th century and standardized private travel; the first model cost $825 (or about $1,200 less than other cars at the time), and the price dropped to $300 by the 1920s (or about $4,000 in 2021's currency).

Not everyone in Eastport had a car or a bicycle, of course, and horses continued to be the primary means of transportation and farm labor. It was not uncommon to see workhorses and trailers moving through streets crowned with steeples in a union of the ideals of hard labor and heavenly faith. The Christ Episcopal Church on Key Street is seen here.

A year after being elected in 1909, Pres. William Howard Taft took 10 days to tour up and down the coast of Maine in the presidential yacht USS *Mayflower*. His visit to Eastport included Todd's Head, considered by some to be the most easterly point in the country at the time (though nearby West Quoddy Head in Lubec, once part of Eastport, now holds the honor).

As a deepwater port on the edge of the country, Eastport was chosen as the recipient of several massive Civil War–era ships that needed to be disposed of. Built in 1867, USS *Franklin* was once the flagship of the United States' European squadron; in 1915, it was sold. Between 1901 and 1920, five such ships were burned in Eastport's Broad Cove.

Few fisheries are as associated with Maine today as that of the lobster, but this is mainly a modern development. Prior to the arrival of Europeans, the Wabanaki people avoided the shellfish, preferring clams, mussels, and sea urchins instead. The Victorians, too, had little to do with it, as demonstrated by its price at 11¢ a pound (compared to 53¢ for the same weight in baked beans). As people became more mobile and traveled to the coast, lobster became known as an exquisite delicacy, and the New England lobster fishery was born. In Eastport, herring continued to consume the attention of hundreds of workers who labored to dry it, smoke it, filet it, or can it. In the 1920s, however, the powerful companies running the canneries created a buyer's trust that limited the economic independence of the area's fishermen.

The workers at a sardine factory in Eastport carry expressions born from difficult work and long hours. On the right, a young boy holds up a knife he commonly used to process the herring. All of the knives used by children and women alike during this time period were the same size—and intended for adult male hands, contributing to numerous injuries. The factories were fraught with unregulated, hazardous conditions, with Maine having only a single factory inspector responsible for patrolling the entire state in 1910. The factories required vast amounts of fuel due to the energy needed to solder the tops to the sardine cans en masse. The air in any given factory would have been thick with charcoal, lead, and tobacco fumes at one point, giving way to kerosene as Eastport became one of the latter's top consumers in the country.

Every step of the sardine canning process was initially contained within the factories. Making the cans was tough work. Tin plates were sheared into strips to create sides along with being shaped for the top and bottom of the can. The tin was then rimmed, bent, and soldered together. Of all the positions within the industry, the sealer was the most sought, as it involved the skillful soldering shut of the filled can. A busted can during shipping would have degraded the reputation of the brand, and so only those with the finest metalworking proficiency were thus employed. One sealer in Lubec, H.B. Thayer, set a local (if not global) record by sealing 1,700 cans in nine hours. Between Eastport and Lubec, 1,616 tons of lead were used for this purpose in 1899 alone.

Eastport's proximity to the ocean may have contributed to its bustling fisheries, but the wealth of the land was in its lumber. Among those who worked the lumber trade was Chin Ferris of South End. Ferris operated a wood yard near the railroad tracks (across the street from Sid Ferris's shoe store) and delivered wood locally with his teamster wagon.

While the high tides of Eastport typically prevented the harbor from freezing—which, combined with its deep waters, made it an ideal port in many ways—Eastport was not immune to the occasional ferocity of winter. Hard freezes with temperatures that dipped lower than –20°F and sometimes held there were met with sheets of ice flowing through the bay.

Built in 1898, the Eastport Grammar School provided a basic education for thousands of children over the course of its existence on Boynton Street. It remained a staple of the community until it caught fire in the 1970s; in its place, the Rowland B. French Health Center was built and remains today. After graduating from grammar school, students moved on to the nearby Boynton High School—at least until 1917, when the school was closed and an imposing modern high school was constructed from brick on High Street. Named after its late benefactor Jesse G. Shead, Shead Memorial High School was widely celebrated upon its gift to the city. This iteration was in use until 1980, when it was demolished and the existing high school was built in its place.

Basketball is the dominant sport of Eastport, consuming the energy of students and adults alike. The sport's origins are close to home, as the reputed first basketball court in the world was constructed in nearby St. Stephen, New Brunswick. Outside of the high school teams and their frequent statewide victories, the adults' Lobsters team was a formidable bunch in the region for decades.

Baseball, as the American evolution of the stick-and-ball game type, has enjoyed a long history in the country and in Eastport itself. Played since at least the late 1700s, baseball provided a reason for families of all ages to spend an afternoon enjoying the sunshine in the abundant outdoors. In Eastport, not far from Shead Memorial High School, spectators line the cliffs of Battery Hill to watch games between local teams.

Football was later to the game, growing in national popularity only toward the end of the 1800s. As attending high school became more and more common for young men in the 20th century, competitive contact sports were developed. These rough-and-tumble players of 1922 are wearing minimal protection against either injury or the weather.

Eastport's northern latitude has graced it with many a cold winter, even as the ocean provides a steady temperature buffer to the fierce inland winds. Fans of winter recreation are rarely disappointed, and hockey and ice skating are common sports on area ponds (such as this one at the corner of High and Adams Streets). Hockey grew in popularity after its introduction in 1875.

Entertainment abounded in Eastport at the turn of the 20th century. Acme Theater opened in 1908 on Water Street as the easternmost theater in the United States. It showed still films and hosted traveling performers for the community's enjoyment until burning down a few decades later. The building housed Oscar Brown's pool hall downstairs and the fraternal Order of the Red Men Hall upstairs; it was replaced by the Wilbor Theater. Among the acts that performed around the community was the Eastport Cornet Band. Brass bands evolved from the wind and string bands of the 18th and 19th centuries as new valve technology (paired with better manufacturing methods) put brass instruments in the hands of musicians all over the world. After the 1920s, audiences in the United States began to prefer a combination of brass and woodwind instruments (such as flutes) in their bands.

Balloons were the joy of the late Victorian world, representing the potential to leave the earthly realm and take to the heavens like blossoms on the wind. Enterprising entertainers would travel across the country to hold balloon ascensions, while famous locales such as the Indianapolis Speedway held balloon races beginning in 1909. Circuses were similarly popular amongst the various communities they stopped in along the railway routes, offering North Americans the opportunity to see exotic animals from Africa, Asia, South America, and Europe. The animals contributed a sense of wonder to the already sensational human performers who showed their skills on trapezes and high wires and delighted the audience as jugglers and clowns. Circuses and events such as this were held at Battery Field. The water tower in the background marks the grounds near the current Shead Memorial High School.

As the country became more mobile and ventured to its farthest reaches, thematic local fairs became increasingly widespread. The Fish Fair seen in this photograph was held in 1916 on September 6 and 7 in celebration of the herring economy. During the time, the city's harbor was continually filled with ships powered by steam or sail during the herring's fishing season as crews worked to haul in catch after catch. Attractions such as "The 6 Legge PolyMooZuke" advertised during the fair spoke to the high appeal of sensationalism during the time period. On the right, modern technology was in full display in the form of an electric photo booth that could produce developed and printed photographs in two hours. Behind it sits the armored cruiser USS *North Carolina*, available for touring. The Fish Fair has endured throughout the decades in various forms and today is held at the end of the summer as the Eastport Salmon and Seafood Festival.

The year 1928 was the biggest Old Home Week to date, featuring an abundance of ball games, parades, and activities. It was not uncommon for the port to attract multiple military ships for the Fourth of July, including one year when there were several destroyers, a flagship, and a supply ship. This shot from 1934 shows USS *Mississippi* (one of the finest ships in the Navy at the time) at anchor near the harbor.

For the Passamaquoddy of the era, parades and fairs were an opportunity to show their solidarity as a tribe. The push to assimilate the Passamaquoddy through the means of cultural genocide continued with the founding of residential schools throughout the northern states. Through the first half of the 20th century, children from Sipayik were sent to the schools and forbidden to practice their language or beliefs.

Perhaps the biggest dream associated with the Eastport region was one held by the developers of the Passmaquoddy Tidal Power Project. The massive tides and unique basin shape of the Passamaquoddy Bay were ideal characteristics for a scheme that would funnel the bay's water through a tidal dam and produce electricity. Engineer Dexter P. Cooper devised the original project in the 1920s after brainstorming collaborative possibilities between the United States and Canada. In 1926, Canada approved the project, agreeing to the dams on the condition that their construction begin within three years. When that time period lapsed, Canada refused the renew the permissions over concerns that the dams would affect local industries. Cooper instead developed two smaller projects in the region that the United States could complete domestically, and from 1935 to 1936, work was conducted in earnest in Eastport. Among those who supported the project (and the associated jobs) was Pres. Franklin D. Roosevelt, who had spent ample time in the Passamaquoddy Bay region in his youth. The model of the dam exists today in the care of the Border Historical Society.

An estimated 5,000 workers would be needed to complete the Passamaquoddy Tidal Power Project. With no housing available, a new village was planned three miles from the city's center. Quoddy Village was designed and constructed as a model village of the time period; it included 128 houses that fit between one and four families each. The village, built on the old George Rice farm, included a fire station, hospital, mess hall, and three large dormitories to house single workers. At its peak, 1,000 workers lived there, all of whom were new arrivals in the area. The mood around Eastport was jubilant as work progressed on new construction projects on a daily basis. The main dam of the tidal project was started on July 4, 1935, with the Army Corps of Engineers providing planning and guidance throughout.

By July 6, 1936, portions of the Passamaquoddy Tidal Power Project were finished, including the Treat-Dudley Dam seen here. This dam bridged between Treat Island and Dudley Island, located about halfway between Eastport and Lubec. These islands and others hosted trading posts and homes of Revolutionary War survivors, with Treat Island being the home of Col. John Allan, aide to George Washington, and Campobello, the home of Benedict Arnold.

The Quoddy Village Administrative Building was an imposing structure that included its own post office, theater, and library. While this building and the barracks have since burned down, the "temporary" housing that was constructed for the workers still remains and is inhabited by Quoddy Village residents to this day.

In Eastport proper, the early Victorian houses that provided most of the housing on the island were occasionally repurposed to serve the goals of the federal government. Seen here in 1936, the Joseph Bucknam House (once the home of dry goods and millinery merchant Joseph Bucknam) operated as a medical dispensary. The Grand Army of the Republic (GAR) building, originally the vestry of the North Baptist Church before becoming the hall for the military veterans of the Civil War, was a secondary administration building for the Army's engineers. The GAR was established on January 10, 1868, and its members would have faded to recorded memory by this time. The GAR building is currently being restored under the care of the Tides Institute & Museum of Art to continue its service to the community as a testament to its rich history.

In July 1936, after a federal review determined that the Passamaquoddy Tidal Power Project would be too costly to pursue, the project was shut down. An ambitious project, it remains the only significant tidal power proposal the United States has ever attempted. The Army continued to monitor the geological and tidal forces in the region, making note of the substantial degree of erosion the islands are subject to.

While the Army Corps of Engineers was no longer staffing and occupying Quoddy Village, the buildings and the facilities were still heavily used by other organizations and the Eastport community. From 1937 to 1943, it was owned by the National Youth Association, and a total of 8,000 "city youths" received training in technical skills as a result. Dances that permitted the invitation of local youths were always a popular attraction.

In the late 1920s, downtown Eastport was a bustling place with personal automobiles filling the streets. Acme Theater is visible in this view of Water Street, along with copious advertisements for the Downie Bros. Circus. This circus was the first to travel by automotive, moving from community to community via a convoy that included 38 trucks.

These men are standing outside of typical contemporary businesses just before the increasingly gift-centered Christmas holiday. On the left is Baron N. Andrews's stationery and bookstore, which would have enjoyed a boom in business in this era due to the interest in postcards. On the right is O.H. Brown's tobacco and periodicals store. This building was located at the bottom of Washington Street directly across from the post office.

Along with owning one of the sardine factories and the shoe factory, the Holmes family was infamous (along with many local businessmen) for their successful smuggling operations. Charles Capen, who continued the tinning operations of G.W. Capen's plant, was notorious for smuggling tin from Canada. Despite the best efforts of customs officials, little could be done to deter smuggling through the porous border.

Over in Johnson's Cove (by Kendall's Head), the Emery brothers operated a fish and whale oil processing plant that produced a steady amount of fish products throughout the year. Roscoe C. Emery was a partner of the fish-packing firm; he later became owner and editor of the *Eastport Sentinel* from 1914 to 1946, mayor of Eastport from 1928 to 1931 and 1935 to 1936, and a member of the Maine State Senate.

Eastport's natural beauty, combined with its coastal fisheries, have long made it appealing to artists in every medium. In the 1920s, among the many artists who came here was renowned landscape artist William Starkweather. Starkweather visited the area repeatedly as he honed his skill at capturing color and perspective. By the 1930s, the number of artists inhabiting and visiting Eastport had swelled, fed in part by an influx from a New York City–based summer art school operating in the area from 1927 to 1936. The era symbolized a time when public interest in depicting the uniqueness of the country in art was higher than ever. On the federal level, President Roosevelt developed a variety of New Deal programs that encouraged artists, writers, and musicians to embrace their calling as a means of enlivening the economy on the heels of the Great Depression.

Marion McPartland, a nurse and widow who moved to Eastport from Massachusetts in the 1930s, acquired the C.E. Capen House (once owned by the vice president of the Eastport Savings Bank) at 23 Boynton Street in 1940. McPartland purchased the home for a singular purpose: to build a private hospital to better serve the needs of the community. She would name it Eastport Memorial Hospital, including the "memorial" portion in honor of her deceased husband and daughter. In 1944, she joined the war effort and sold the hospital to the Eastport Hospital Corporation. It would continue operating as a private hospital until 1981, when it was closed due to the increasing number of restrictive federal regulations. It reopened as Eastport Memorial Nursing Home and continues to provide a valuable service to the families of the island.

The number of cars in the United States rose from 8,000 in 1900 to 23 million by 1930. As the mid-century dawned in Eastport, so too did car parks, campgrounds, and motor inns, each welcoming weary travelers who were on adventures that tested the limits of the growing highway infrastructure across the nation. Among the lodgings that cropped up on the outer banks of Eastport's shore were the Harris Point Shore Cabins and Seaview Campgrounds (both of which continue to operate today). These seaside cabins provided visitors from inland with a rare and rustic experience on the shores of Passamaquoddy Bay. The photograph below shows a woman enjoying Cony Park in Broad Cove with a schooner relic behind her. (Below, courtesy of the St. Croix Historical Society.)

New areas of Eastport became cleared for development, including Redoubt Hill. The large houses were built as part of the Passamaquoddy Tidal Power Project and represented modern model homes with full amenities. Unlike the temporary housing for the workers in Quoddy Village, the houses at Redoubt Hill were built for senior administrators and engineers and were made to be permanent homes.

Back on the eastern shore of the island, small, ramshackle houses along the cliffs stood watch over the ceaseless sway of the tides. While protecting against the elements was difficult without insulation, early summer in Eastport is often a treat for the senses. Taken on July 10, 1940, next to Prince's Cove, this photograph offers a clear glimpse into life in the seaside community.

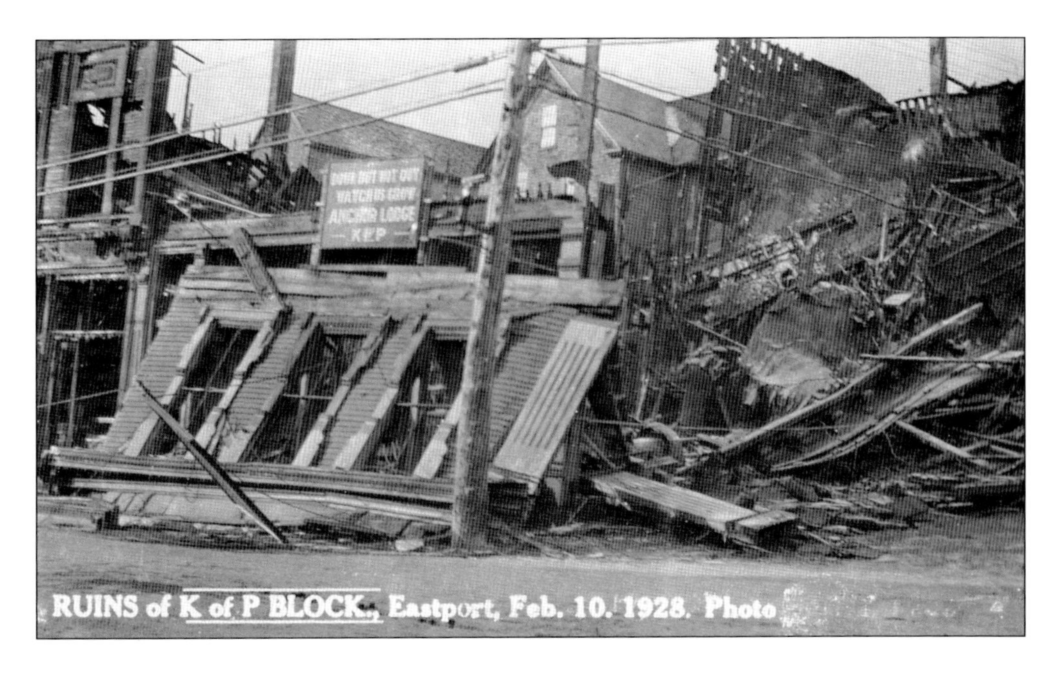

While relatively free of major disasters since the Great Fire of 1886, Eastport has not been without its share of spot catastrophes. In early February 1928, the building housing the Knights of Pythias anchor lodge burned to the ground; the organization rallied and continues its efforts to support the community today. Train wrecks were uncommon but disastrous due to the limited working area afforded to machinery and crews. Boats were sometimes employed to salvage goods and, when necessary, to remove debris. On the financial front, the Great Depression was enough to cause Eastport to declare bankruptcy in 1937, joining many other municipalities in the state who had done the same. The cost of assisting residents and businesses through the sparse time period had drained the city's coffers, and it remained under state control until 1943.

Since 1937, the National Youth Association had been garnering support to have an airport built in Eastport to enable it to train its charges in aircraft flying. The impetus grew as Quoddy Village began hosting a contingent of Women Ordnance Workers Corps trainees who were learning to build and maintain aircraft. In December 1942, Eastport Municipal Airport officially opened.

Along with Moose Island, Deer Island, and the Wolves, Dog Island is a small companion not far from Harris Point. A modest lighthouse, seen here in 1952, warns vessels of the outcropping. At night, it is but one of many lights that shine from near Eastport's shore, collectively brightening the island as it awaits each day's dawning.

Five

Tides of Adaptation

As Eastport looked around from the crest of the mid-century, it was clear that the world was changing rapidly. World War II had repositioned the United States on the global stage, and cities like Eastport were undergoing a continual metamorphosis as their long-stalwart extractive industries faded away.

During the war, Eastport's sardine industry thrived as the federal government mandated a diet of canned sardines for its troops. After the war, however, fishermen were noting a sharply decreased herring stock, and the truth of the sardine industry's end on the horizon was becoming abundantly clear. Recognizing the need to pivot, Eastport readily took on new opportunities that presented themselves in the latter half of the 20th century.

In rapidly developing Quoddy Village, change was constant; it served as a hub of activity for the region for decades. World War II had prompted the facilities in the village to shift away from the National Youth Association's purposes and into military production skills in its new role as Navy Seabee base Camp Lee-Stephenson. Men and women alike learned how to work foundries, use machine tools, and manipulate aviation sheet metal.

After the war, Quoddy Village was briefly considered to serve as a transitional home for 25,000 displaced Jews and their families in a plan proposed by New York entrepreneur Frank Cohen in 1947. The village was under the control of the federal War Assets Administration, however, and it set a price tag at $300,000—too high for Eastport and the plan's proponents to afford.

In the late 1950s, a modern causeway was built through Sipayik, creating a route that would become critical for Eastport's land freight purposes in the future. Grocery stores such as A&P and IGA provided a centralized place for residents to buy food, and appliance stores offered the most revolutionary appliances.

Even with Eastport's efforts to adapt, its population was steadily shrinking. From 1900 to 1950, the number of residents decreased from 5,311 to 3,123, creating a dramatic shift in the city's momentum. Those that remained poured seemingly boundless energy into the community's survival, and the nurtured threads of creativity and industry proved enough to sustain it.

Among those who adapted most over the second half of the 20th century were the Passamaquoddy. The tribe was aided by its continuing line of successful, influential chiefs who were bringing greater awareness to indigenous pride. Chief Joseph S. Nicholas, who was periodically chief from 1925 to 1948 at Sipayik and later became the model for Indian motorcycles, was one such example of this.

In downtown Eastport, the sardine canneries stretched far out from the shore on spindly legs. The year 1952 was the peak one for Maine's sardine industry, with 6,000 workers in 52 factories around the state; afterward, the catches decreased one after another. In 1983, the last factory in Eastport closed. (Courtesy of Library of Congress.)

Among the fish factories that remained through the mid-20th century was B.H. Wilson, which was later purchased and run by Jim Warren. Wayne Wilcox, who was in his teens when he joined the ranks of multigenerational fish factory workers in his family in the 1970s, recalls how the packing at the facility (and in other factories) was always done by the women due to their dexterity and speed. Wilcox's grandmother, Alice Flagg, was the fastest packer at B.H. Wilson and traveled along to help train the workers when a new factory was opened in Grand Manan, New Brunswick. Many of the workers on both sides of the border at this time were migratory. On the American side, they worked in the fish factories through the warmer months before traveling down to Connecticut and Massachusetts to work in other industries for the winter.

Elsewhere in Eastport, new industries provided much-needed jobs through the end of the 20th century. The Mearl Corporation, founded in Eastport in 1933 by New Yorkers Harry E. Mattin and Francis Earle and Eastport resident Burton G. Turner, employed as many as 250 workers at a time in manufacturing pearl essence from herring scales until eventually closing in 2007.

In 1954, Guilford Mills moved into Quoddy Village, providing jobs for 175 employees in its 60,000-square-foot facility. It was a new branch of the Guilford Woolen Company, which began in 1865 in Guilford, Maine. In 1978, the factory suffered a fire; it was rebuilt but later fell into disuse.

Sleek new automobiles and pleasure boats were common sights in 1957 Eastport, providing stark contrast with the utilitarian buildings lining the waterfront from the previous century. Visible at left on Water Street is Havey & Wilson, a pharmacy that would continue serving the community for decades (including from a second location on Washington Street). On the right is the back of the Wilbor Theater.

The collapse of the sardine industry contributed to a decrease in rail freight for Maine Central Railroad, and in 1978, the Eastport Branch (seen here on Sea Street) was fully abandoned. The building behind it was originally the American Can Factory; later, women (known as the "Sea Street Strippers") would peel the skin off herring to prepare it for packaging.

In 1955, Winifred French moved to Eastport with her husband, Rowland, who would go on to help found the Eastport Health Care Center (later named after him). In 1968, dismayed by the number of community newspapers closing their doors in the area, Winifred started the *Quoddy Tides*, assisted by a small staff. Among them were Inez Segien, holding up the paper as Winifred points, and John Pike Grady, peering into the typesetting machine at Sterling Lambert's print shop in Deer Island, New Brunswick. In 1979, Winifred was named Journalist of the Year by the Maine Press Association, and in 2018, son Edward—who is continuing his mother's legacy as editor—accepted her induction into the Maine Press Association Hall of Fame. The *Quoddy Tides* plays an active role in keeping the cross-border community informed in an otherwise relatively sparse news landscape.

Widely considered to be the city's cultural ambassador for generations, red-bearded John Pike Grady is seen here outside of one of Eastport's enduring eateries. In 1919, Winfield Cummings purchased a photographer's traveling studio and converted it to a lunch counter parked at Bank Square. In 1923, new owners Nelson Watts and Ralph Colwell named it the Wa-Co, and in 1971, the structure was rebuilt in its present form.

In the 1960s, the completion of the Eastport breakwater enabled vessels of up to 700 feet to dock. Seeing the opportunity, Alexander Brown opened a food stand centered on hot dogs cooked in peanut oil. Later, Rose and Afton Whelpley ran it, followed by Rose Marie Lingley (seen here) in 1972. Rosie's remains a community staple under the care of daughter Paula Bouchard, who took over in 2000. (Courtesy of Edward French.)

The breadth of the breakwater, which paired with Wadsworth's pier to create the mouth of the sheltered harbor, is apparent in this photograph of Eastport from 1972. At the base of the breakwater on the left, Rosie's hot dog stand is visible as a small white building, while the larger white structure to the left of Rosie's is the office of the *Quoddy Tides*. Every day, scenes of the waterfront unfolded in this nexus of activity, from ships being loaded and unloaded to fishing vessels bringing in the daily catch to whale-watching boats filling up with curious sightseers. Up the hill, to the right of the photograph's center, the water tower marks the site of Shead Memorial High School, not far from where circuses and community fairs were once held. Nearby, the crumbling shell of the powder house represents the last remnants of Fort Sullivan in its original location.

Through the late 20th century, massive ships from Europe and Asia were escorted by humble tugboats to the breakwater. The foreign cargo has caused its share of ecological issues, including the arrival of the European fire ant (*Myrmica rubra*), first officially reported in 1952 (although some recall it coming in on the railroad long before). Without competition, the ant has staked claims to various areas around the island and beyond. (Courtesy of Tom McLaughlin.)

The network of islands surrounding the Passamaquoddy Bay—regardless of their nationality—have long been tightly linked by water transportation. Ferry Landing, located next to the shuttered J.W. Beardsley's Sons factory (closed in 1958 and demolished in 1960), once served as a hub for travelers looking to get directly to other communities without having to drive around the coast first.

The mailboat *Rex IV* operated through the 1960s transporting important communications and packages between the islands and from up and down the coast. Based in St. Andrews, it traveled around the bay collecting mailbags (and occasional passengers) from each town and leaving others for distribution. In the center of the photograph on top of the hill sits a house built where Fort Sullivan once was.

Along Water Street in the 1970s, after the surge and retraction of the prosperity wrought by the sardine industry, businesses steadily closed and buildings fell into disrepair. City infrastructure became increasingly weathered as the funds to fix roads and sidewalks were harder and harder to find.

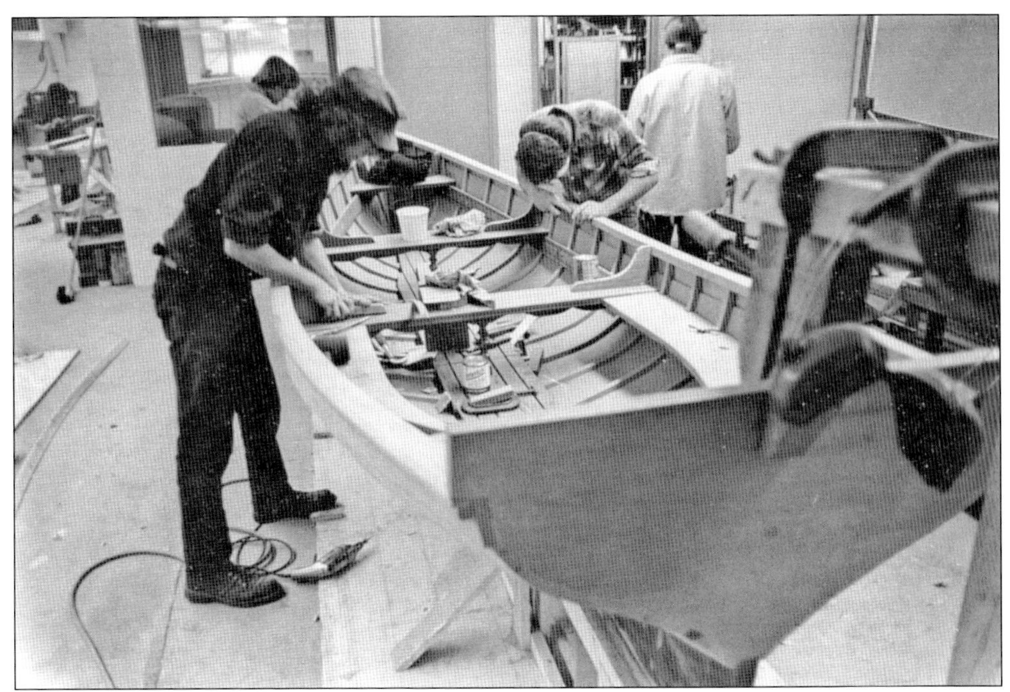

The maritime tradition of Eastport is one of the community's greatest assets. In 1969, the one-of-a-kind Boat School was founded to provide students from all over the country with hands-on knowledge of New England boatbuilding techniques. Founders Ernie Brierley, Doug Dodge, Fred Beal, the Calder brothers, Junior Miller, John Pike Grady, and Peter Pierce passionately fueled the school's growth and successfully navigated its many obstacles for decades.

The 1970s saw an expansion of Eastport's artistically minded inhabitants, many of whom were seeking refuge from the high rents of New York City. Artists of all persuasions purchased homes in the area, along with raising $2,000 to purchase the former Lyons sardine factory at Prince's Cove. The factory was renamed Sunspot and hosted dozens of creative studios.

Along with economic challenges, Eastport's greatest threat in the 20th century has been its weather. While the average winter can provide any number of calamities, some winters are especially fierce. The Great Blizzard of 1952 and that of 1978 were notable in the amount of snow, wind, and ice they delivered, requiring residents to band together to see to their daily needs.

In 1976, the Groundhog Day Gale struck Eastport with a fury not seen since the Great Saxby Gale. Wooden buildings across the downtown, including the four-story Cherry Block and the Wadsworth store at Central Wharf, were demolished, and the Sunspot was swept completely out to sea.

Regardless of the source of the challenge, the Eastport community has endured it, with those who survive often made stronger for it. The S.L. Wadsworth store continues its operations in downtown Eastport today, selling a myriad of supplies for construction, fishing, repair, and virtually everything an island resident might need.

Year after year, the streets of Eastport swell with the arrival of former residents and tourists who come to see the largest Fourth of July festival in the state. Old Home Week is a symbol of local, regional, and national pride and features a parade that regularly attracts thousands.

New ventures utilizing the natural resources of the area added fuel to Eastport's momentum, including the first commercial seaweed farm in North America. Coastal Plantations International, established in 1991 through the efforts of Eastport resident and aquaculturist Steve Crawford (foreground), funding partner Ike Levine (background) of New York, and Shep Erhart of Franklin, employed dozens of workers in cultivating nori. The seeding and drying operations took place in a former sardine factory. (Author's collection.)

Aquaculture and shipping are the undisputed backbones of Eastport's economy. Salmon pens and the Estes Head Shipping Terminal, built in 1998, are visible from Shackford Head, which itself was formed by the receding Laurentide ice sheet and turned into a state park by community advocates in 1989. Salmon farming has been a staple industry from the 1970s onward, including Ocean Products, Connor's, and today's Cooke Aquaculture. (Courtesy of Robin Hadlock Seeley, Maine Island Photo.)

Six

THE ENDURING PORT

From its origin as Moose Island, the granite outcropping that would become known as Eastport has held a special place in the memory of those who have inhabited it across time. For the Wabanaki, "People of the Dawn," it was a place to share and thrive in the cycle of seasons; for the French and English it represented a haven of natural riches and opportunities beyond their imaginings; and for Americans, it was the wild coastal frontier.

Generation after generation, the human residents of Eastport have watched the happenings in the world from one of the first places to see the sunrise in the United States each day. The wild elements surrounding the island, including the powerful tides and Old Sow, one of the five largest whirlpools in the world, lend to ample creativity and innovation amongst a community that is nonetheless dwindling due to age and a lack of economic opportunity.

The long story of Eastport, however, is one of adaptation. Those who remain to face the city's challenges are typically passionate advocates for its culture and its history, and a number of organizations are now working together to envision a brighter future based, in part, on the city's rich past.

As a young generation of tourists explores the American landscape, Eastport stands out as an unusual gem, and more and more people are discovering the city through its increasingly renowned festivals. Remote workers in a variety of industries are relocating to the island, drawn by its natural beauty and relatively low cost of living, while artists continue to arrive for the same reason. Soon, day flights to Boston will bring a new level of accessibility to the area.

Like the rest of Down East Maine, Eastport fits the description of rural. Being perched on the edge of the continental United States, however, has always given the city a worldly flavor enriched by swirling currents. With each incoming tide, fresh nutrients arrive, and so the coastal community continues.

In its mature years, Eastport is perhaps best symbolized by the humble mushrooms that grow from beneath its fallen trees. The bulk of the community—its culture, its history, those who have come before—lies underground, providing a sturdy foundation for fruiting bodies to surface, cycle after cycle, generation after generation. (Author's collection.)

Meanwhile, in Passamaquoddy Bay, the diverse ecosystem faces its own set of challenges and opportunities. Seals, once hunted to near extinction, now flourish under legal protection, creating problems for the lobster industry and attracting more sharks to the bay. Rockweed, which provides a vital marine nursery habitat, is facing the opposite problem and is now controversially harvested as a cash crop for overseas markets. (Courtesy of Don Dunbar.)

At the heart of Eastport is its children, youth who grow up in neighborhoods two centuries old and look out upon the ceaselessly changing harbor. Like those who have come before them, the city's youngest generation is tempered by the experience of coming to age in a remote region where amenities are not always easy to come by. Unlike their predecessors, however, Eastport's youth benefit from an exceptional level of technological connectivity. Whereas physical distance once built cushions of time against changes happening in the nation's urban centers, the suddenness of the internet has brought Eastport fully into the high-proximity cloud. For those in the midst of childhood, the task becomes balancing the nearly infinite potential of digital exploration with the ample joys and sensory experiences to be found growing up in coastal Maine. (Courtesy of Edward French.)

Eastport's past few decades have been defined by a willingness to look for opportunities. When television reality show *Murder in Small Town X* chose Eastport to represent the fictional town of Sunrise in 2001, this statue was built by the film crew to add "authenticity" to the fishing city. The community was endeared to the statue, with many seeing a resemblance to now renowned John Pike Grady, and decided to keep it. Just one week after the airing of the finale, 35-year-old Angel Juarbe of New York—the winner of the $250,000 reality show contest—was among the firefighters who rushed in to save the victims of the falling towers during 9/11. Sadly, Juarbe perished in the effort, and the *Fisherman's Statue* now stands in part as a memorial to him and the fallen firefighters who met their fate attempting to rescue survivors following the terrorist attacks. (Courtesy of Robin Hadlock Seeley, Maine Island Photo.)

Keeping a watchful eye on the harbor is *Nerida*, a bronze mermaid statue built by metalworking artist Richard Klyver in 2015 as part of a community-sponsored effort to amplify the artistic footprint downtown. Klyver, a New York native, relocated to Eastport in 1974 after spending three years in Kenya. The playfully smiling sea nymph was specially requested by donors as an homage to the eternal mysteries of the sea. (Courtesy of Don Dunbar.)

On this rare early evening in June, the center of Eastport is empty as the buildings providing its skeleton quietly await the return of residents from their festivities elsewhere. Though aged, the buildings are being preserved; 30 properties in the downtown commercial district alone are part of the National Register of Historic Places. (Courtesy of Don Dunbar.)

Shipping is fueling Eastport's economic growth in the new millennium, with the port itself being the city's largest employer. The Estes Head berth is 64 feet deep at low tide, which pairs with its relative proximity to Europe to make it ideal for shipping commodities from the States. In 2010, it became the only terminal in New England approved to ship live cattle internationally, enabling it to send 500 pregnant cows to Turkey. (Courtesy of Edward French.)

A significant portion of Eastport's working population continues to make their living as seasonal workers harvesting natural resources. The work is difficult and sometimes life-threatening, with fishermen and mudflat diggers (collecting clams and bloodworms) alike facing numerous unpredictable environmental challenges and hard physical labor. (Courtesy of Don Dunbar.)

Viewing Eastport from the breakwater today, the ships it protects are a combination of working vessels and pleasure boats, lined up side by side above waters still teeming with marine life. The downtown buildings are in a continual state of repair or maintenance as the community works to reverse the impact of decades and the elements. Eastport's history and its culture are being recognized as among its finest assets, leading to a steady progression of art gallery openings and community preservation meetings throughout the warmer months. The post office and former customs building looms large over incoming vessels, providing a stately reminder of the port's commercial maritime tradition. To the right and out of frame, Moose Island Marine, run by Boat School alum Dean Pike, stands ready to provide boaters and fishermen with any supplies they may need. Much like the granite it rests upon, the diversity of its composition—fishermen, merchants, teachers, artists, historians, and so on—lends to Eastport's strength. (Courtesy of John Jackson.)

Among those who have created new enterprises from old materials is Ross Furman. After growing up in Eastport, he returned as an adult in 1987 to claim and renovate a 1790 farmhouse owned by Capt. Jacob Lincoln. The fully renewed property now provides unique lodging opportunities for travelers under the name Rossport by the Sea. (Courtesy of Jeff Gonzalez.)

Eastport's Border Historical Society owns a former barracks building of Fort Sullivan, now relocated to Washington Street, and hopes to restore the structure for public viewing. Downtown, the society shares a space with Quoddy Crafts, where it hosts a model of the Passamaquoddy Tidal Power Project. In 2020, the society placed its collections, including a host of military artifacts and models, with the Tides Institute & Museum of Art. (Author's collection.)

One of the biggest attractions each year is the Eastport Pirate Festival, started in 2005 and drawing thousands of enthusiastic "pirates" to join in revelry and a host of events and activities. Residents and visitors of all ages participate in the fun, from learning how to swashbuckle to watching fire eaters to drinking copious amounts of grog at the evening soirees. Most of the city transforms to adopt the pirate theme for the occasion, providing a slightly surreal twist on the port's robust smuggling history. Included with its biggest supports are Don Dunbar (below, front left) and partner Kathleen (below, front right), who operate the Eastern Maine Images gallery in their off-pirate hours to showcase Dunbar's award-winning photography. (Above, author's collection; below, courtesy of Don Dunbar.)

Though winter's chill drives the snowbirds away and suppresses many of the community's activities, there are some who remain committed to celebrating together throughout the year. The annual Sardine Drop is the easternmost New Year's Eve gathering of its type, bringing warmth and cheer to help welcome in the slow return of longer days. (Courtesy of Don Dunbar.)

Artists from all over the world (including Lena Schmid of Brooklyn, seen here) take time to find inspiration and create in Eastport as part of the StudioWorks residency program. Along with the Sardine Drop, StudioWorks is one of many programs or activities organized by the Tides Institute & Museum of Art. Founded in 2002, the museum serves a vital role in community preservation, including as the steward of eight historic buildings.

While the community is small, Eastport's teachers, parents, and volunteers go above and beyond to provide engaging, experiential activities for the youth. Students have learned how to grow food, build mechanisms and prototypes, and sculpt metal under the tutelage of various instructors over the years. The student and community-centered WSHD-LP radio station, meanwhile, provides a unique service to the island and surrounding communities. At the Eastport Arts Center, children from Eastport and neighboring schools have performed a number of productions as part of the Passamaquoddy Bay Symphony Orchestra's Music for Children program. In the photograph below, a musical performance of Brundibár, written by Jewish prisoners for their captors during the Holocaust, featured a valuable lesson about coming together to overthrow recurring bullies. The children's voices were accompanied by the professionally trained orchestra (with Greg Biss conducting). (Above, courtesy of Edward French; below, author's collection.)

As it has been for generations, Old Home Week remains the joy of the island's youth—and those who fondly remember being young. Eastport's dedication to its Fourth of July activities has made it the largest celebration in the state. The penny scramble and codfish relay are two activities that have enjoyed long decades of participation from energetic and eager youth looking to gain a day of fame. Other events include a greasy pole competition, a rubber ducky race, and a doll carriage parade. In the daytime, vendors and food stalls line the streets, while in the evening, the torchlit Callithumpian parade and fireworks provide ample entertainment. Up at the airport, the annual blueberry pancake breakfast fills the air with the sweet aroma of the locally harvested fruit. (Both, courtesy of Edward French.)

From distant beginning to the immediacy of the present, the Passamaquoddy tribe has persevered through generations of turmoil and oppression. Fortunately, the end of the 20th century marked significant changes for the tribe, including the Maine Indian Settlement Claims Act of 1980. It enabled the tribe to take steps toward economic independence and improve the lives of tribal members with better infrastructure, but it was far from a silver bullet in resolving the tribe's embattled history. Opening the door further, Maine formed a Truth and Reconciliation Commission (the first of its kind in the United States) for the Wabanaki people within its borders, and the findings have promoted a new understanding of the enduring generational trauma. For many tribe members, sharing the stories is restorative, and Passamaquoddy culture is finding strong and willing vessels in the tribe's youngest generations as a result. Each year at Sipayik, the Passamaquoddy tribe honors its ancestors and celebrates its living connections in a public festival at the beginning of August. (Author's collection.)

Nestled deep in the heart of Eastport is a resounding truth: While the surface may change with the fury of tides and the relentlessness of time, the community's foundation remains firm on pillars of solidarity, companionship, and resourcefulness. This piece, *Stillness in Change*, is one by Elizabeth Ostrander, a famed artist in the community who relocated from New York in 1971 as part of the back to the land movement. Her work in various media incorporates feminine energy with natural wonder and mystery, making it particularly appealing for coastal audiences looking for meaningful connections in the wild environment around them. Ostrander is a founding member of the Eastport Gallery, which itself has become an important venue for many artists in the area since its 1985 inception. This work, along with numerous others, has been shown at the Eastport Arts Center. (Author's collection.)

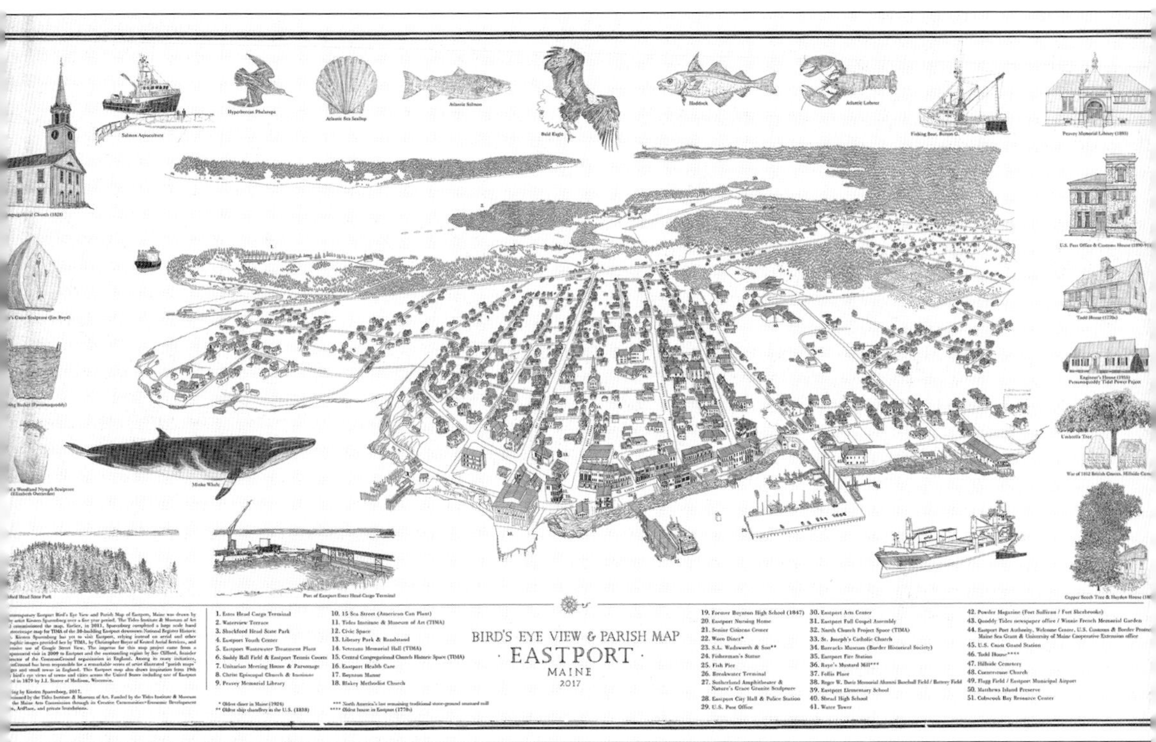

The Eastport of today (shown here in a bird's-eye view by Kirsten Sparenborg) has a distinctly different profile than the city of 100 years ago. Rather than a chaotic array of dozens of wharves, a single large breakwater and neatly spaced piers project into the Passamaquoddy Bay to organize the flow of marine traffic. Houses built centuries before still line the streets, but they are joined by newer structures and modern infrastructure. Massive cargo vessels ply the waters en route to Estes Head, while whales and dolphins are regularly visible traversing the waterways between islands. History thousands of years deep shows but a crest on the surface in the present, yet every development throughout time rests squarely on the city's rich past. The sturdiness of that foundation is, in part, what makes Eastport a community embodied by its willingness to continue to lift its eyes toward each new dawn, even as it faces the uncertainty of the future.

INDEX

About the Tides Institute & Museum of Art

Founded in 2002 in Eastport, Maine, the Tides Institute & Museum of Art is located on the Atlantic coast at the US/Canada border. This boundary location shapes its perspective on the region and the broader world: through its initiatives and programs to foster new innovative and cross-sector works; through its wide-ranging collections, education, and preservation efforts; and through its partnerships and endeavors to strengthen the region's economic prosperity, vitality and wider connections. Visit the institute online at www.tidesinstitute.org.